C000072237

Cultivating Arctic Landscapes

Cultivating Arctic Landscapes

Knowing and Managing Animals in the Circumpolar North

Edited by David G. Anderson and Mark Nuttall

Berghahn Books
New York • Oxford

First published in 2004 by

Berghahn Books
www.berghahnbooks.com

©2004
David G. Anderson and Mark Nuttall

Library of Congress Cataloging-in-Publication Data

Cultivating Arctic landscapes : knowing and managing animals in the circumpolar North / edited by David G. Anderson and Mark Nuttall.
    p. cm.
    Includes bibliographical references and index.
    ISBN 1-57181-574-0 (alk. paper)
    1. Eskimos--Domestic animals. 2. Eskimos--Hunting. 3. Sami (European people)--Domestic animals. 4. Sami (European people)--Hunting.
5. Domestic animals--Arctic regions. 6. Pastoral systems--Arctic regions.
7. Reindeer herding--Arctic regions. 8. Human-animal relationships--Arctic regions. I. Anderson, David G. (David George), 1965- II. Nuttall, Mark.

E99.E7C85 2003
636'.00911'3--dc22

                                                                    2003063590

British Library Cataloguing in Publication Data

A catalogue record for this book is available from
the British Library

ISBN 1-57181-574-0 hardback

Contents

List of Figures

Foreword

My closest neighbour, when I was living among the Skolt Saami of northeastern Finland in 1971–72, was Piera Porsanger. Piera was not himself from a Skolt family. His ancestors had inhabited the area for many generations before the Skolts were resettled there after losing their original homelands beyond the postwar border with the Soviet Union. However, he had married the daughter of one of the new arrivals, and by the time of my fieldwork he had a large and bustling family. They were not well off, and Piera always wore the expression of a man worn down by the perpetual worry of having so many mouths to feed from a small and uncertain income. Fishing was poor, and though wise in the ways of reindeer, Piera had not come out well from the upheavals that had afflicted herding during the previous decade. A combination of severe overgrazing, a series of disastrous winters and attempts to use snowmobiles to round up the now scattered and depleted herds had meant that a substantial proportion of each year's crop of calves had gone unmarked, only to be snapped up the following year in auctions by the handful of younger men who had spearheaded the new techniques of snowmobile herding and thereby cornered reindeer mustering operations to themselves. Like many men of his generation, Piera had seen his herds melt away, and he was keeping afloat only thanks to his brothers-in-law, who were among the most active of the new enthusiasts for mechanised herding. Yet behind his care-worn look lay a twinkle that gleamed through the thick lenses of the glasses he always wore to correct his myopia.

For Piera was a philosopher. He thought too much for his own good, people would say, just as his irrepressibly nosy wife Maria gossiped too much for hers. That was why he had lost most of his reindeer. Indeed Piera was one of wisest and most knowledgeable men I have ever met, fluent in five languages (three kinds of Saami, Finnish and Norwegian), and immensely curious about the ways of the world. Despite his appalling eyesight, he was an acute observer of everything that was going on around him. He appeared continually and genuinely astonished, and yet nothing really took him by surprise. Astonishment, for him, was a way of being, revealing an openness to the world that, by the same token, left him peculiarly vulnerable to its vicissitudes. Someone unfamiliar with this way of being might interpret it as a mark of timidity or even weakness. I had often wondered myself why Piera had allowed himself to be trampled on by all and sundry, with no more than his usual rejoinder of quizzical resignation. Only later did I begin to realise that his approach to life was one widely shared by Saami people of his generation and older. In their attitude of unsurprised astonishment, which for many outsiders indicates a lack of intellectual rigour and moral fortitude, lay the very source of their strength, resilience and wisdom. But it is an attitude that has earned them little respect from those who assume that the way to know the world is not by opening oneself up to it, but rather by 'capturing' it within the meshes of a grid of concepts and categories. For astonishment has been banished from the procedures of conceptually driven, rational inquiry. It is inimical to science. Yet scientists are forever being surprised by the apparent failure of the world to conform to their calculated predictions. They have even turned surprise into a principle of creative advance, converting an accumulation of errors into a record of consistent progress.

As I was reading the chapters that make up this volume, I kept thinking of Piera Porsanger and what he taught me. All around the circumpolar North there are people like him, people of extraordinary insight whose lives have been crushed underfoot by the power of a system of knowledge comprised by the history of its mistakes. Nowadays that system, as if to make up for past failures, is increasingly concerned to harness the knowledge of native inhabitants, but it can do so only on its own terms – that is, as classified information. The result, in some regions of the North, has been an unseemly scramble to collect the 'traditional ecological knowledge' of older generations before it is too late. Noone to my knowledge has asked Piera for his TEK, but if they did, I am sure they would be met with the same quizzical glance, issuing from those twinkling eyes behind the spectacles, that always seem to throw the question back at the questioner: 'Well, what do *you* know?' As I found during my fieldwork, and as many other ethnographers of the North have found both before and since, knowledge for native inhabitants consists not of information that can be transmitted, but of wisdom that one grows into. It was assumed that having grown up in my own society, I must know *something*, but to know anything of what my hosts knew I would have a whole lot more growing up to do. It is no good asking other people, I would have to find out for myself. Though they could not provide me with ready-made knowledge, they could at least provide opportunities for me to learn.

Like most wise men of the North, Piera was a great storyteller. He made up many of his stories himself, and they were by turns poignant and comic. I shall always remember one such story, about a reindeer on its way to find fresh pastures, following a migration route that it and its kind had followed for many years. As it peacefully meandered along familiar tracks, it was astonished – but not surprised – to come up against a newly constructed barrier fence that cut straight across its path. Piera concluded his story with the voice of the reindeer: '*Where the devil do I go from here?*' The deer's question, I think, goes to the heart of the issues raised by this book. It does so in three ways.

First, it reminds us of what is obvious to northern native people, that reindeer – like most other animals – are sentient and intelligent beings with points of view of their own. Piera was, of course, putting words into the reindeer's mouth, but this was only his way of describing what a real reindeer would actually feel on encountering the equally real fence. Unable to keep on going, it feels frustrated and disoriented. Now according to the canons of official science, to attribute feelings to animals is to commit the cardinal sin of anthropomorphism, of treating the animal as if it were human. It is a condition of scientific inquiry that the objective world of nature, including all nonhuman animals, should be closed off from the world of society to which human beings alone are admitted as rational and sentient subjects. Yet it is precisely by this closure that the scientist is prevented from developing knowledge about animals in the way that native inhabitants do – that is, by opening up to them just as one would to fellow humans, and by making their experience one's own.

Secondly, the dilemma of the reindeer in Piera's story forces us to reflect on why the new fence had been built across its path in the first place. The story refers to a fence that had, in fact, been recently built along a boundary between adjacent herding districts, to prevent herds from one district from wandering onto the pastures of the other. There is nothing new about the construction of reindeer fences. They have been used for generations as devices to funnel the movements of deer towards round-up enclosures, and before that – in the days of reindeer hunting – towards pitfalls, snares or ambush positions. But in all such cases the fence, along with the contours and features of the landscape, has formed an integral part of a *trap*. The trap is a kind of story-in-reverse, embodying in its construction an account of the movement and behaviour of the target animal or herd as it proceeds towards its goal. What was relatively new, at the time of my first fieldwork in Lapland, was the use of fences to *enclose* entire pasture districts. These fences serve not to funnel but to block the movements of animals. The rationale behind them is that the enclosure of pastures allows for better regulation of animal numbers in relation to the availability of grazing. The point of Piera's story, however, is that the animal is not just a number, nor is it in its nature to stay put on a bounded block of territory. The construction of boundary fences indicates an obsession with compartmentalisation and control that flies in the face of any sensitive understanding of the animals, and to which they cannot be expected to respond positively.

Thirdly, the story is an allegory for the situation in which northern native people increasingly find themselves. It has never been their aim to remain bound to fixed routines, forever reenacting the practices of their ancestors. 'Traditional society', in that sense, has never existed in the North. Rather, people have aimed to *keep on going*, through improvisation and adjustment in response to a close perceptual monitoring of ever-changing environmental conditions. Time and again, however, they now find their path ahead blocked by imposed regulations, restricting their access to, and use of, the land and its fauna. Like the reindeer in the story, they experience frustration and disorientation. This frustration is compounded by the insistence of those in authority that the restrictions are for their own good, or for the good of future generations. Thus northern people are caught in an impasse in which they are told that the only way to continue hunting is to stop hunting, so as to allow the herds to pick up; and that the only way to continue herding is to stop herding, so as to allow the pastures to recover.

It is undeniable that right across the northern circumpolar region, native inhabitants face enormous challenges in keeping life going. These challenges are richly documented in the chapters of this book. But we should perhaps resist the temptation to lay the blame too readily at the door of arrogant or uncomprehending scientists, patronising and overly bureaucratic managerial regimes, or distant states that co-opt both science and management to their authoritarian and centralising objectives. Whether the numbers of caribou in northern North America really declined in the mid-twentieth century, as many wildlife biologists maintained at the time, can probably never be known with any degree of certainty, nor can we ever know for sure whether – if they *did* decline – native hunters bore any responsibility for this. But it is not impossible that they did. Nor can we automatically absolve reindeer herders of any responsibility for overgrazed pastures. The scenario of the tragedy of the commons may indeed project a characteristically Western rationality, far removed from the realities of life in the forests and tundra. Nevertheless it was a truism among the Saami herders I knew that security lies in numbers, and that when everyone strives to increase the size of their herds serious overgrazing of common pastures inevitably results. Everyone was aware that this was happening. Native people are not 'original ecologists', guided by a tradition that, unbeknown to them, causes them to act in ways that place ecosystemic sustainability before their own interests.

Nor, on the other hand, did scientists, bureaucrats and officials arrive from outer space, fuelled by an unworldly desire to run this planet according to rational principles of sustainable management to the general discomfort of its indigenous populations. For they, too, are the sons and daughters of inhabitants, people who have had their own connections to the land of one sort or another. Most often, albeit a few generations back, these connections have been established through the practices of farming. For me it was an eye-opener to work among Finnish reindeer herdsmen whose forebears had been peasant farmers and forestry workers. Here I found them thinking of their herds in terms of the yield of meat from the land; the size of a round-up would be estimated in carcasses, and reindeer owners were negotiating collective agreements among themselves in order to ensure that limits for individual herd sizes were not exceeded and that owners killed enough females and calves in each year to keep overall numbers within the capacity of an enclosed territory. It is true that they were doing all this in accordance with principles that are now enshrined in Finnish reindeer management law. But the law itself is based on conventions and agreements that evolved in the agricultural settlements of northern Finland, over a period of some three hundred years, specifically in order to deal with issues of cooperation, scheduling and conflicts of interest between farming, forestry and reindeer herding. In following these conventions, reindeer owners did not feel themselves under the heavy hand of an interfering state. Of course the state interfered in other ways, such as in its enforcement of measures to protect bears and wolves, about which they complained vociferously. But when it came to basic principles of reindeer herd management, these were felt to be deeply rooted in the practicalities of farming, not in the abstractions of science.

It would be over twenty years before I saw Piera Porsanger again. By then, a new express highway had been built through the Skolt settlement area, running through to the coast of northern Norway. Unlike the old dirt track, the highway turns its back on the community and defies the contours of the landscape. Driving along it, one would not realise that people lived there. The tiny cabin in which Piera and Maria somehow managed to cram their enormous family was deserted. The old track had veered so close to the cabin that one corner almost stuck out into it, and this proximity had enabled Maria to keep close tabs on all the comings and goings in the community. But the new highway bypassed the cabin. I found Piera in a neighbour's house. He was sitting in the back room reading a newspaper, as always keeping up with world events. He was astonished to see me, but not in the least surprised. He began to talk as though we had last seen each other only yesterday, as so indeed it seemed also to me.

This book tells of the trials and tribulations, and of the major challenges and minor triumphs, of the people of the North whose lives revolve around reindeer and caribou. But in reading it, do not ever forget how astonishing is the northern environment for those who live there. We need to hold on to that astonishment, and to celebrate it. And we need to resist the inclination to turn the North into a world of spectacular surprise. Surprise exists only for those who have forgotten how to be astonished at ordinary things, who have grown so used to control and predictability that they depend on the unexpected to assure them that events are taking place and that history is being made. That is how the West has made a history for the North, through the catalogue of its magnificent failures, above all in predicting the numbers and behaviour of terrestrial and marine fauna. The animals, however, are never surprised, though they are often astonished. They do not expect the world to conform to expectations, and nor do the people for whom they are the staff of life. We can learn from them.

Tim Ingold
University of Aberdeen
June 2003

Acknowledgements

The idea for assembling a book comparing the meaning of caribou or reindeer for people in various northern places originated during a late evening discussion between Peter Usher, Ivar Bjørklund and David Anderson in Ivar's kitchen in Tromsø, following the plenary sessions of the Tenth International Ungulate Conference in 1999. We developed the idea further during a special panel session at the Twelfth Inuit Studies Conference in Aberdeen in 2000 (which was convened by Mark Nuttall). David Anderson, Patty Gray, Murielle Nagy, Natasha Thorpe and Peter Usher presented papers on how caribou or reindeer (or their absence) became images tightly connected to state power. Following the conference, we widened the plan of the book to include a focus on knowledge about caribou in the contributions of Hugh Beach, Craig Campbell, Julie Cruikshank and Frank Sejerson.

We would like to thank the Social Sciences and Research Council of Canada and the Sustainable Forestry Management National Centre of Excellence for research grants, which funded the field and library research of David Anderson, Craig Campbell and Rob Wishart. We gratefully acknowledge the support of the Dower fund of the Principal of Aberdeen University, which allowed Natasha Thorpe to participate in the Inuit Studies Conference.

Special thanks is due to Erin Payne, who helped us to format and assemble the book manuscript through funds provided by the Forest People's Programme.

List of Abbreviations

ABWLP	Advisory Board on Wildlife Protection
ACIA	Arctic Climate Impact Assessment
AEPS	Arctic Environmental Protection Strategy
AMAP	Arctic Monitoring and Assessment Program
BCMPC	Bathurst Caribou Management Planning Committee
CAFF	Conservation of Arctic Flora and Fauna
CITES	Convention on International Trade in Endangered Species
COSEWIC	Committee on the Status of Endangered Species in Canada
CWS	Canadian Wildlife Service
DIA	Department of Indian Affairs
DIAND	Department of Indian and Northern Development (Canada)
GHL	General Hunting Licence
GN	Government of Nunavut
GNWT	Government of the Northwest Territories
HBC	Hudson's Bay Company
ICC	Inuit Circumpolar Conference
IFA	Inuvialuit Final Agreement
IPS	Indigenous Peoples' Secretariat
IQ	*Inuit Qaujimajatuqangit* lit. 'the Inuit way of doing things'
IQTF	IQ Task Force
ITQ	Individual transferable quotas
IUCN	International Union for the Conservation of Nature
IWC	International Whaling Commission
IWGIA	International Working Group for Indigenous Affairs
KANUKOKA	Association of Municipalities in Greenland
KHS	Kitikmeot Heritage Society
KNAPK	National Association of Fishermen and Hunters in Greenland
NAB	Northern Administration Branch, Department of Northern Affairs and National Resources (Canada)
NAC	National Archives of Canada
NLCA	Nunavut Land Claims Agreement
NRTA	Natural Resources Transfer Agreement
NSDC	Nunavut Social Development Council
NTKS	Naonayaotit Traditional Knowledge Study
NWMB	Nunavut Wildlife Management Board
NWT	Northwest Territories
NWTGO	Northwest Territories Game Ordinance
RAIPON	Russian Association of Indigenous Peoples of the North

RC	Roman Catholic
RCMP	Royal Canadian Mounted Police
SEPA	Swedish Environmental Protection Agency
TEK	Traditional Ecological Knowledge
TEKMS	Traditional Ecological Knowledge Management Systems
TNP	Tuktu and Nogak Project (Canada)
UNESCO	United Nations Educational, Scientific and Cultural Organisation

1

Reindeer, Caribou and 'Fairy Stories' of State Power

David G. Anderson

Northern places are often spoken of in extreme, uncompromising terms. For many they are understood to be harsh, cold, remote and romantically challenging. These extreme metaphors are not innocent rhetorical flourishes. As Lakoff and Johnson (1980) have shown us, metaphors frame concepts in such a way as to shape the way people respond to them. The history of northern peoples is full of such misunderstandings. What is seen as the 'desolate' Arctic has become a dumping ground for the steaming artefacts of the Cold War (Osherenko and Young 1989). In the idiom of international law, northern territories are seen as *terra nullius* – empty frontiers wide open for settlement and appropriation of mineral wealth (Richardson 1993a). The anthropological canon, at times, has been no less innocent for its tradition of placing Arctic hunter-gatherers in test-case studies at what Harvey Feit (1994) has evocatively identified as 'the absolute zero of human culture'. In an ironic reversal of terms, the 'fragile' Arctic ecosystem spurs urban environmentalists to protect it with nature reserves and management regimes, often by forcibly evicting the people who use the land and thus impacting the environment in a different way (Lynge 1995b; Catton 1997; Spence 1999). Stefansson (1921) and Berger (1977) prominently employed concepts such as 'the Friendly Arctic' and 'Arctic Homeland' in order to fight these metaphors with words which force us to focus upon the people living in the North. However, even here the human element of the phrase stands in an unexpected and defiant contrast to the coldness of the geographic term.

The chapters in this collection came together gradually through a series of conversations where many of us expressed frustration over the way that the usual metaphors of the North steal the meanings from the words we use. In order to counter this tendency we assembled a volume where the North could be understood in a way recognisable to people who live there. Instead of focusing on the idea of the cold or harsh North, all chapters examine knowledge, power and tradition as reflected through the example of one central northern protagonist: the reindeer or caribou.[1] Each of us has conducted long-term research in a variety of northern communities, often in different languages. In every case, whether we are speaking Greenlandic, Russian, Saami or French the image of the reindeer or caribou stands out as a focal symbol for the warmth, vibrancy and generosity of the land. To draw two examples from my own research, Siberian Evenki reindeer herders are endlessly curious about the pelage and the *taste* of North American Porcupine caribou that they have seen in my photographs. The eyes of Gwich'in hunters light up when one describes how hunters in Siberia approach migratory caribou swiftly and silently with harnessed white reindeer on a white birch reindeer sled.

This collection tries to reconceive northern landscapes not as remote frontiers to civilisation but instead as cultivated places. In using this term we are trying to capture three aspects of the term 'cultivation'.

The first is that northern hunters have very special ways of knowing the world which often tie together moral and technical, as well as rational-economic, forms of action. In making this argument we ask readers not to think of cultivated action as a 'high' cultural practice of recording and inscribing behaviour on texts. Instead, as Ingold notes (2000: Ch. 5), it should be thought of as a broader type of personal cultivation employing memory, listening, watching and participating. The stories in this collection give examples of how Inuvialuit or Chukchi live together with reindeer or caribou to create a unique web of relationships in a cultivated environment.

The second way we understand the metaphor of cultivation is in an active mood. Our research focuses upon the way that the northern biosphere is radically shaped by the actions of people. Hunters create the environment through selective hunting, flexible or inflexible kinship institutions, or through the weapons they use (as they are often reminded by wildlife managers). Scientists and administrators also shape the biosphere with their licensing regimes, monitoring practices and the special weaponry of 'wildlife management' itself. While northern places may not be cultivated with plough and spade, we argue that deliberate action, communicated through how people engage animals, forms a major anthropogenetic effect on the world around us.

Third and perhaps counterintuitively, we would like to communicate to our readers the sheer vibrancy and sociality of northern landscapes. Many of us at interdisciplinary conferences have been forced to shyly admit that the 'sample size' of our research is only a few families or at best a community of one or two thousand people. However, to understand the social dynamics of northern places we insist that we must also strive to understand them as densely populated places – a place populated with creatures rich in intentionality, history and connection to the people around them. We would like readers to feel, without a sense of irony, what the North feels like when it is experienced as populated with caribou.

This collection is unique for its international focus. Each of the chapters builds upon the similarities in human stories in all the regions of the circumpolar North, whether in Eurasia, Greenland or North America, or irrespective of the fact that political regimes are labelled social democratic, socialist or capitalist. One sobering similarity in all regions of the North is that administrative management style has a similar ecological footprint on human–animal relationships, be it in the paternalist mode of the postwar welfare state, or the hierarchical model of Soviet state socialism. The Cold War may have divided the North along formal ideological lines, but the chapters in this book show that it had a remarkable effect in homogenising relationships between people and animals on either side of the polar frontline. If rational state management of lands and peoples can recognised as a form of cultivation, this volume shows how it was conceived as a pervasive monoculture with little tolerance for local styles and initiatives. By pointing out the costs of this standardisation of practice we hope to draw attention to the value and beauty of places where people manage relationships with animals directly.

Stories as Focal Knowledge

The Evenki village of Surinda has a long pedigree in the cultivation of reindeer. At the height of Soviet power it was declared a *plemzavod* – a special state agency charged with producing pure-bred stock which could be sent out around Russia to improve the form of socialist reindeer. The start of my doctoral fieldwork was the end of this distinguished period in the history of this community. In 1992 the *plemzavod* was turned into a joint-stock society. By 1997 the numbers of domestic reindeer held by the shareholders had plummeted from 10,000 head to a mere 1,000, while the forests surrounding the village rustled with the sounds of the world's first feral caribou herd born of privatisation. Today I often meet with refugees from Surinda – intelligent, ambitious men and women who are trying to build lives for their families in the new urban centres of post-Soviet Russia. Much like people the world over displaced from evocative rural contexts, they often speak about their lost their reindeer, describe their 'faces' and names, and tease each other in the rude, bawdy way that only Soviet *surindtsy* can do.

One person who remains in Surinda is Nikolai Gavrilov, who in 1992 was the brigadier of the Number 4 reindeer brigade. In 2001 Gavrilov was still the head of a group of contract herders who kept a group of 500 reindeer belonging to twenty different families. His life and that of his son, is devoted to reindeer such that in their world the categories of person and of reindeer merge together. One of my clearest memories of Gavrilov is an aggressive meeting between him and the school headmaster in the village boarding school, in the days before reindeer culture became unravelled by the policies dictated by the International Monetary Fund. Gavrilov had been summoned because his son, Sasha, was skipping classes and failing at core subjects such as Russian language, mathematics and history. Gavrilov, typically, refracted the blame back on the school headmaster: "What do you mean he needs discipline? He and I lead (*upravlaiut*) one thousand head of reindeer! You can't even manage (*ne spravlaiete*) three hundred children!" Gavrilov stomped out scoffing at the superfluousness of this rural school. The headmaster similarly chuckled to me about what to do about a man who thinks that training a reindeer is of the same order as training a child.

I often think about this fundamentally absurd exchange between two specialist trainers who held different philosophies on how to cultivate persons. In comparing these philosophies I will not here refer to the large literature on human and animal personhood, which is excellently summarised by Cruikshank and Wishart in this volume. Suffice it to say that I think it would be unfair to be distracted by the obvious physical differences between human children and reindeer. One may stand on two feet and the other on four but – at least according to Gavrilov – both should find an appropriate place in the Surinda taiga. This exchange challenges us to imagine how learnt behaviour should be properly embodied in the local environment, whether we are speaking about humans or animals. Unfortunately, the history of state-led development in the North gives us many negative examples of this challenge.

One of the sad ironies of Soviet rural development policy is that though Siberian rural people were urged to build socialism with a disciplined approach to education, technology and a rational division of labour, the skills that they were given were those that aimed to preadapt them to live as proletarians in a Moscow suburb. The same could be said of the reindeer, who were corralled, counted and cultivated to be more disciplined than their 'wild' cousins. In each case, the skills given to both children and reindeer could not survive without the structures imported and subsidised from outside. If the former

Soviet Union ever had a vision of socialism, it was a vision that looked towards cities and to the south (see Gray, this volume). It is a sad irony that the visions of Scandinavian and Canadian social engineers were similar (Usher, Beach and Bjørklund, this volume).

Gavrilov's comments to the headmaster ask us to imagine a different way of cultivating northern landscapes wherein children, hopefully, understood mathematics, but also where people and reindeer understood one another in a way that enskilled and empowered each other to live a good life in that place. This is seemingly an ever more romantic vision, at least for post-Soviet Siberia. They say that even today the children flown into the Tura central boarding school from Surinda are the strongest of all children in the entire district in maths and sciences (and one can only admire the headmaster, who, like the brigadier, stubbornly sticks to his post). However, these well-trained schoolchildren grew up in a place that looks more and more like a remote outpost of a crumbling industrial civilisation where neither the state, the market, nor the forest provides a more predictable livelihood than one of the focal points of Soviet reindeer husbandry. It is also said that today there are increasing numbers of children who have dropped out of school and who live full-time in the forest. Indeed I have met members of the post-Soviet generation – young men of nine or ten years of age – living in remote areas of Siberia, who spend so much time on the land that they scarcely understand Russian. Here we see the necessary corollary of a failed industrial model. Not only are some children kept away from the cultivating influence of the forest or tundra, but the state is no longer able to reach out to cultivate its own children.

It would seem that cultivation can only be useful, or beautiful, if it is calibrated to its context, much like the way that Borgmann (1984) describes how appropriate techniques create 'focal things' (Strong and Higgs 2000). Borgmann's prime metaphor is the difference between a wood fire and a central heating system, where the former unites members of a family in what is somewhat romantically described as a harmonious division of labour. The latter is obviously a metaphor for alienation. I am not so sure that northerners would voluntarily surrender the idea of central heating so quickly, at least not without large numbers of junior affines to do the work of cutting and splitting firewood. However, if we consider that one's attention can be focused not only by something as mesmerising and warm as a fire, but also by something as clever as a good story, we can catch a thread which helps us understand northern ideals of cultivation better. Whether we speak of the Canadian, Scandinavian, or Russian North, clear, exacting and knowledgeable speech is in all these places valued as a mark of wisdom, respect and maturity. Obviously good speeches are not limited in their repertoire to the themes of reindeer and caribou, but caribou are certainly a widespread subject in good stories, as the contributors to this volume demonstrate. The main point of contention is how true stories can be best focused on their context.

One of the main misunderstandings which pervades scholarship is that truthful knowledge of northern places can be 'captured'. Cruikshank and Thorpe in this collection present us with multiple examples of the damage and violence done to northern visions when the statements of knowledgeable people, often 'elders', are taped, transcribed, codified, labelled and filed. Cruikshank suggests that the production of TEK (Traditional Ecological Knowledge) changes knowledge into something very different from the way that it pervades evocative speech contexts. Thorpe argues that we can only understand Inuit approaches to focal knowledge as *pitquhiit* – that is, a set of traditions and knowledge related (in this case) to caribou. Both authors acknowledge that it is often profitable and politically necessary for people all across the North to package their knowledge in order to fight specific battles in courtrooms and boardrooms. However, they sound the same tragic warning note, as with the Surinda story: that the more the focus of human activity leaves the forest and moves to certain authorised places, the more the ecology of knowing itself changes. When the critical analyses of Cruikshank and Thorpe are compared to the story of the industrialisation of reindeer knowledge in Bjørklund's or Gray's chapters, one obtains an even clearer picture of the way that state systems have promoted a monoculture from one side of the Arctic to the other.

How can one contextualise knowledge so that the very frame in which one organises words does not transform the topic of our conversation? One of the more hopeful techniques also comes from within the growing universe of studies of TEK. Both Thorpe and Nagy (this volume) present excellent renditions of how the words of knowledgeable 'elders' can be set off from the text in an authoritative manner, giving these statements primacy, immediacy and a very seductive sense of legitimacy. In Nagy's case, she weaves together a detective's story of how caribou were replaced by another ungulate – muskoxen – despite the warning cries of hunters who saw this biological disaster looming. Instead of telling the story in a fixed narrative, the elders give us not only a biological account of why caribou left the people, but also a critical analysis of the history of biological science. Through their words we see how management concepts and measurement practices led to a regulatory regime that was unfriendly to caribou – and led to their exodus from the human world.

It must be noted that authoritative accounts such as these, which put local elders in the same positions of authority once occupied by scientists, share a similar weakness to orthodox science studies. The readers must assume that authority flows from age and experience and that the author is best placed to adjudicate their legitimacy. Sejerson's account (this volume) of battles between wildlife biologists and local Greenlandic hunters gives us a different, critical perspective on these issues. He argues that Greenland does not have 'elders' but only 'old people' and links the status economy in authoritative statements in North America to the very special colonial situation of the North American nation-states. Although Gray does not go into depth concerning the authoritative knowledge of Chukchi elders, I can confirm that in Central Siberia age does not necessarily lend the right to make unquestionable statements. In post-Soviet Siberia, in a dramatic contrast to the North American Arctic, degrees and higher education are often the badges that make individuals from the native sparse nationalities members of the 'native intellegentsia'. Even here, there seems to be some evidence of cultural diffusion of the principle of elderness along with development assistance from international nongovernmental organisations. Very recent Russian-language publications, not to mention the political statements of aboriginal rights groups, use the term often (Krupnik 2000b; Anderson 2001). Sejerson's chapter might be understood as a call to research the history of the use of the term 'elder' in Canada, where it seems to have a specific history in relationship to a continent-wide aboriginal rights movement (Cruikshank, pers.com.).

One way to resolve the question of authority might be to examine how genre and style affect the way that knowledge is presented. Most powerful speeches occur in special contexts that are often understood historically or intuitively. In this case the onus is on the writer to build the context which makes first-hand accounts authoritative (Rushforth 1992; 1994). In this volume, the work by Thorpe, Wishart and Nagy gives us clear examples of how threats to livelihoods, such as the taking of children off the land, or the very *absence* of caribou give good cause for talking even more intensively about caribou. Cruikshank underscores the responsibility of the listener/reader to gather enough information in order to understand statements competently. A recent collection of essays suggests that ethnography itself offers a good example of how to simultaneously weave evocative descriptions of context to discussions of a moral or political nature (Anderson and Berglund 2003).

However, to bring this discussion back to the case at hand, can a good story help to recultivate the local landscape around Surinda? This is a sobering thought, especially when it is brought up in the context of the severe disempowerment of rural communities in Siberia. It would seem that the only way to answer this is to go to the very root of northern stories themselves. Good stories touch upon important issues that breathe meaning into our being on this earth. They are, as Julie Cruikshank (1998: 24) observes, a way of 'bridging social fractures that threaten to fragment human relationships'. To use an important philosophical category from all areas of the circumpolar North, they are *real* in the sense that wholesome food is 'real' (Fienup-Riordan 1990: Ch. 2) or a competant Nganasan hunter is 'real' (Gracheva 1983). Real stories, although they remind listeners about their relationships to knowledgeable people around them, are rarely supported by authoritative bibliographies. Instead their authority usually comes from the co-presence of a person who has spent many years of living well in what everyone can recognise as difficult circumstances. The stories of Gavrilov and the school headmaster are real each in their own way. They each give children, and reindeer, a choice between two trajectories. According to the headmaster, excelling in formal scholarship allows children to leave Surinda and the North, as quickly and effortlessly as possible. According to the brigadier, excelling in living with reindeer both helps the reindeer prosper and makes life for local people more entertaining and more secure. In writing about both stories, I hope that I can bring home the message to readers in Europe and North America of the tremendous cost of economic restructuring on the fabric of local relationships.

If there is a moral to these stories, it would be that the deafening monologues of urban-based industrial development are not good stories. Tim Ingold (2002), in a recent chapter on the history of northern ethnography, describes development-talk as a 'ready-completed heritage'. In his view they lack the evocativeness of local stories which have no beginning and no end and thus cycle onwards like any other form of life. In thinking about good stories in this way we no longer focus upon their literal message but upon the way they help listeners understand their places in the world.

Recent ethnographic work in the North, building on Basso (1990), has been investigating the link between stories as events which frame meaning and spatial ideas of 'sites' and 'places' which frame meaning equally well. Emma Wilson's (2002) recent ethnography of environmental movements on Sakhalin Island, Siberia gives pride of place to the way that engaged political movements create spaces where Nivkhs and other local people can live. Muehlebach (2001) also focuses upon the important sense of 'place' cultivated by delegates at the International Working Group on Indigenous Affairs, who try to transform the legalistic environment of an international committee into an indigenous camp. The chapters in this volume also look at the way that talk about reindeer and caribou help us construct sites which allow relationships between people and caribou unfold in a 'never-ending' manner. Here readers can identify the ways that local communities make Inuitness out of board meetings or Gwich'inness out of environmental management. In this volume we argue that stories about cultivating *Rangifer* offer an opportunity for discussing how to move culture forward.

Contemplating Reindeer and Caribou

Northerners are not the only ones who culturally appropriate northern reindeer and caribou. If the image of caribou or reindeer occupies any conceptual space at all in the minds of southern, urban populations, it is certainly as an object of contemplation. This is also a powerful way of knowing animals, but it is qualitatively different from placing animals in a focal relationship to one's own lifestyle. In this mode of thought we are not troubled to accept dialogue from the animals themselves, but instead use them to as models to illustrate aesthetic relationships. Examples of contemplation range from the pretty but clumsily harnessed reindeer of Father Christmas on postcards, to the ever-expanding industry of conservation which seeks to alternately preserve or enhance the wildness of certain stocks. In both examples, the practical relationship between people and reindeer is muted (or, in the case of parks, often outlawed). The best reindeer to contemplate are often wild reindeer.

As Wishart observes in this volume, the idea of that a northern animal could be wild at all is a horrifying thought to northerners. What is underrepresented in most popular and scientific literature are the pains which northern hunters take to cultivate populations of animals which seem as though they should be beyond their direct control. Wild reindeer, or caribou, are prosaic creatures since they seem to respond to very deep and accurate instincts which propel them great distances along 'traditional' migration routes. Understanding the mysteries of these migrations is one of the great challenges to modern wildlife biology (Anderson 2000b). Protecting these populations seems to be one of the most powerful symbols of modern-day wildlife management. The philosophies behind both practices take us deep into stereotypes of what is special and unique about scientific models.

As Gail Osherenko (1988) remarked a long time ago, wildlife management is not so much about managing animals as it is about managing people. Despite the fact that Inuit, Evenki or Gwich'ins insist that they can influence the presence of caribou by sharing meat, or by dressing properly, most professional wildlife managers assert that one can only guarantee the presence of animals by restraining the predatory actions of people. Thinking about the relationship between people and deer leads us to examine some of our assumptions about behaviour in general. In this case, the relationship between people and deer advocated by formal models in wildlife biology is marked by a great fear that 'irrational' use could bring about their extinction. Local hunters are thought to possess great insight, but are thought to employ it towards short-sighted ends. The way state managers contemplate deer lead them to take special measures: the separation of intuition and tradition from responsible action. In the rational and protected environment of human-caribou relationships, all action is mediated by hypotheses, the authority lent by higher education and statistical accounts of encounters rather than first-hand experience.

The historical chapters in this collection give rich examples of the operation of this ideology. Campbell's work on the history of the notion of 'wanton slaughter' in the Canadian Arctic gives a clear example of the operation of the mythical imagination on material that is portrayed as strictly scientific. Here, as with Milton Freeman's work (1989; 1992), we see a very human story of how the citation and re-citation of credible opinion quickly stifles the real issues on the ground. This leads to the tragic conclusion where the correct interpretation of events takes second place to the imperative to save the caribou from the savages. In Usher's chapter we are given a very detailed notion of the evocative concept of the 'crisis' in wildlife management (see also Kofinas 1998). Here the notion of a crisis, Usher argues, was created in order to re-embed northern administrative economies following changes to the postwar global economy. It is interesting to reflect how similar pressures led to a very different type of social engineering in Soviet Siberia and in social democratic Norway. Here, as with Campbell's genealogy, hearsay plays a great role in the sequence of events. In rebuilding northern economies, the tools which were marshalled were the authoritarian use of subsidy, the restrictions on human interactions with caribou/reindeer, and, perhaps most disturbingly, the displacement of local kinship-led relationships to animals. The chapters by Gray and

Bjørklund in particular document the eclipse of clan and *siida* cultivated spaces by an entirely centralised structure of allocation.

It is tempting to dismiss these historical examples as the mistakes of
sloppy science, but this would discount the central message of these
four chapters. In each case from Canada, to Norway, to Russia, the driving ideology seems to be a competing idiom of cultivation: the necessity for there to be a clear, rational, accountable way of planning
nature's providence. These early examples of audit cultures (Strathern
2000) for animals makes the notion of the superstition and the 'irrational' especially paradoxical. While state managers ridicule the way
that local ritual is employed to summon caribou in a cultured way,
they invest their careers in setting up auditing regimes to bring people's ecological action to conform to their own standards of morally
productive use. Looked upon from the point of view of local hunters,
their efforts are 'irrational' on a much broader scale. The direct impact
of this rational-moral action is often a weakening of the fidelity of animals to traditional migration routes, unexpected jumps and dips in
population structure and breakdown in the reproduction of hunting
traditions within local human communities.

One of the most astounding examples of the application of urban
contemplation comes from Hugh Beach, where he tells us the story of
efforts to cherish the last Swedish wolf. In this account aboriginal
Saami reindeer herders are branded as 'eco-criminals' for worrying
about attempts by the state to artificially reintroduce a 'natural' predator in a space already heavily impacted by roads, management regimes
and other artefacts of industrial audit culture. Draped with mythic
symbolic import, the Swedish wolf is monitored and guarded with
state-of-the art technical gadgetry by functionaries oblivious to the
place in which it lives. The spark for this explosion of contemplative
energy seems to have been the random event of a wolf pair accidentally
crossing a national boundary.

The chapters by Wishart and Nagy give us a more tragic example of
applied wildlife management. Here the tool of 'translocation' was
employed to cultivate Banks Island and the northern Yukon for an
extinct form of megafauna: the late Pleistocene muskox. Pleistocene
translocation is another growth industry in applied circumpolar environmental protection. As Peter Lent (1999) documents, wildlife managers seek to restore past Pleistocene environments thought to have
been depopulated by ancestral humans, armed with state-of-the-art
Clovis-point stone tools. The translocation of muskoxen is part of a
growing international movement involving the reconstruction of landscapes into 'circumpolar Pleistocene parks' of wood bison, muskoxen

and even attempts to reincarnate woolly mammoths (Zimov and Chuprynin 1991; Stone 1998; Guthrie 2001). As Wishart and Nagy document, wildlife managers are motivated by a deep sense of duty to cultivate palaeo-landscapes in order to restore a larger sense of balance in ecosystems. These attempts tend to be practically reinforced by the need to cultivate landscapes that are friendly to eco-tourists. These examples raise the issue of identifying in whose interest a landscape is cultivated. This comes through quite clearly in Nagy's research, where Inuvialuit hunters show how they were conscious that at any time they could reintroduce muskoxen to their island, but chose to manage their home for caribou instead. The dependence of so-called wild populations on the human politics is very clear in these two chapters.

The tools of cultivating northern landscapes are not limited to the authoritarian regimes of wildlife management, the generous application of central subsidies, or to Clovis-point tools or for that matter, repeating automatic weapons. TEK-inspired research resoundly shows the important role of ritualised respect-relationships. This can range from the sober notion of propriety that motivates Gwich'in hunters not to 'bother' animals with unnecessary human presence (Wishart) to Inuit models demanding that meat be shared generously to ensure that animals come back. The stories in Cruikshank's chapter speak much more directly of the cross-over in attentiveness between people and caribou as attuned hunters become caribou and vice versa. The beauty of local ethnographic accounts of caribou is that they remind us that not all aspects of people's relationships to animals can be 'captured' by technically or even morally based arguments. In extreme circumstance, as Cruikshank and Thorpe remind us, setting up 'codes of proper behaviour' distracts us from listening and attending to the place in which we have an interest. These codes can even be wielded as weapons against the hunters from whom they were first distilled.

As with the previous section, these paradoxical examples of so-called wild landscapes cultivated for multiple purposes direct our attention to how we can write about spaces in an idiom of potentials rather than an idiom of templates, codes and rules. The examples in this collection give clear examples of how literary analysis, environmental history and ethnography allow us to know northern ecosystems in a complex way. This certainly does not imply that through knowing them we are directly changing them, but at least we can share different stories which can inform responsible action.

Four Legs Good: Two Legs Bad

In my own research and in editing this book, I have been struck by the degree of convergence in what seems to be a circumpolar fairy tale of the separation of people from animals. As northern folklorists will recognise, such tales are extremely common in the repertoire of first nations as well as Siberian societies. These are 'real stories' which capture a fundamental truth about timeless relationships between living things generally (Scott 1996; Nabokov 2002: Ch. 2). If there is any truth behind these recent examples of the separation between people and animals, it seems to lie in the message that hierarchically organised state structures of any stripe seem to intensify this process rather than alleviate it. They are, using Latour's (1993) analysis, institutions which will not overtly tolerate mixed or hybrid environments but inevitably leave half-built or half-rationalised hybrid contexts in their wake. For the serious comparative scholar of hybrid modernist economies, the circumpolar North provides some of the most disparate regimes knowable. Here we have combinations of socialist and capitalist; urban and rural; shamanic and monotheistic. Yet at the intersection of these unfinished regimes is similar fallout of the not quite successful attempts at power and control. Whether we speak about Alaska or Siberia, we learn about how populations of caribou begin to behave 'wildly' as soon as they come under the 'rational' control of wildlife management regimes. In Scandinavia or Canada, evocative ways of watching and interacting with animals are replaced with hierarchical regimes of controlling and measuring people and animals. In each of these regions, the stories of a focal interrelationship with caribou, or reindeer, are replaced with morally tinged stories of rational use, protection and predictable outcomes.

There is nothing mystical about this convergence of ideals of action and morality in each of these frontiers of the industrialised North. At the end of the Cold War, economic development and the cultivation of livelihoods are coming increasingly under the control of global committees that soberly adjudicate local agendas. This too is a subject of narrative. George Orwell's (1945) 'fairy story' *Animal Farm* cleverly inverts the European practice of using animal stories to bring up children in order to focus the reader's attention on important moral and political points. In this story, as in Native American or Siberian legends, there is no hard and fast division between animals and people. The actions in one sphere bear a direct message for the other, much as the skill of raising children in boarding schools is implicit in the cultivation of a herd of reindeer. Unlike the positive model of this relationship in Cruikshank's chapter about how a boy became a caribou, the 'fairy story' technique is adopted in order to make an ironic statement of a lesson of social power that should be obvious even to a child. Orwell's work is both evocative and time-tested in that it captures and criticises power dynamics on both sides of the circumpolar Cold War front. In doing so it summarises nicely the convergence that each of the authors in this volume has noted in postwar economies, no matter their stripe.

One of the organising themes of this volume is the contrast between people who adore wildlife and those who understand animals as part of a cycle of birth, consumption and death; a cycle which includes human beings. As analyses of the animal rights movement have shown (Wenzel 1991; Dizard 1994; Preece 1999), urban philosophies tend to interprete animals as feeling but atomised individuals who are thought to enjoy rights similar to those of any individual creature in a liberal-democratic scheme. The idea of an anthropomorphised animal right is often portrayed as a right to 'enjoy' life without the threat or pain of death. This stands in stark contrast to a common idea in northern hunting societies that it is immoral not to 'take' animals when they present themselves. For Evenki, or for Inuit, the caribou which shares its body with the hunter and his or her kin is thought to agree to surrender itself for consumption as part of a common struggle to preserve life. Here, there is a notable absence of the hypothesis that life is to be enjoyed by individual creatures. Instead it is thought to be something that neighbouring organisms share with each other. Mutual consumption and predation – hunting – are thought to be part of this process.

The Orwellian irony in the ideology of 'four legs good – two legs bad' was, of course, that though four-legged animals are contemplated and adored, two-legged animals are nonetheless the measure by which all things are valued. By treating animals 'as if' they were (not quite) humans, their unique personhood is denied. The managers of the circumpolar Animal Farm walk upright. Rather than trying to imagine themselves in the place of animals – to think 'as if' they were a caribou – they try to speak authoritatively on behalf the species. In doing so, they hamper others such that they are discouraged from apprehending caribou or reindeer in their own right. As Harvey Feit (1998) reflects, management rules are paternalistic trust relationships set up 'in the name of animals'. Their emphasis is usually on the legitimacy and authority of the speech-act rather than on cultivating a good life with animals. It is this trick of representation – this unrealness – that creates the power imbalance in most management contexts.

In the original work, *Animal Farm*, the faithful horse Mollie was condemned to death for taking joy in wearing the farmer's ribbons. This harsh sentence reflects another common hard principle that animals should remain 'wild'. Like the pigs of Snowball's administration, rational-technical managers also feel that there should not be a mixture of practice and attention between people and animals. The worst examples of this strictness can be found in the historical chapters of Usher and Campbell. This premise once again contrasts universally with local practice across the circumpolar North. Most local ways of perceiving caribou or reindeer involve the postulate that they are sentient beings that are able to hear, judge and react to human thoughts. On Evenki or Saami lands, there is a literal convergence of practice and ideals where especially valued reindeer are dressed with beaded harnesses (a practice discouraged under Soviet reindeer husbandry). In more radical examples, now unfortunately a generation in the past, certain Evenki reindeer that grew up and matured along with human children, were thought to be twins who share the same 'road'. Such people and deer were thought to be able to shoulder the burden of illness from one another, or to use their breath to literally breathe power and vibrancy into each other (Anderson 2001). These reindeer–people twins were bred out of existence by the forced resettlements and boarding schools of high Sovietism wherein reindeer were forced to abandon their clothing and humans were ordered to walk a different road. The mark of the circumpolar Animal Farm is the stricture that people and caribou everywhere must move in separate orbits, not investigating each other's personhood.

The moral of Orwell's fairy tale is that one must be very cautious of the words that are wielded to exhort people to do good deeds for others. Language and metaphor are powerful tools which do shape perception and action. The North as commonly perceived is understood to be a place for desperate action and not a place where one should listen and attend. In writing about places populated with caribou the contributors in this book call on us to imagine how action fits into a wider context – a context where northern animals listen, watch and share their lives with people. Like the conversation between the reindeer brigadier and the school headmaster, northern places should be first of all places where there is continuity of skill and practice between people and reindeer – both are part of the same open-ended story. The contributors to this volume see it as their task to make this possible by directing our attention to the symbols and words that make us imagine the North as a warm, populated place.

Acknowledgements

I would like to thank Julie Cruikshank and the participants at the Inuit Studies Conference Panel (Patty Gray, Natasha Thorpe, Peter Usher, Rob Wishart) for their comments on earlier drafts of this paper.

Notes

1   Despite the fact that *Rangifer tarandus* is a common image in all regions of the North, it is known by radically different terms. In Eurasia both wild and domestic *Rangifer* are called reindeer (and often European biologists use this term when speaking about North America). In North America the Micmac word 'caribou' has become standard. In this volume both terms are in use, depending on the region. In this comparative chapter, when I use either 'reindeer' or 'caribou' I am using the terms to mean *Rangifer tarandus* in the broadest sense (see Banfield 1961a; Anderson 2000b: 159–60).

2 Uses and Abuses of 'Traditional Knowledge': Perspectives from the Yukon Territory

Julie Cruikshank

Since the 1990s, discussions about indigenous knowledge or traditional ecological knowledge have become internationalised, both in scholarly debates such as those emerging from environmental anthropology, and as part of daily discussions in indigenous communities where anthropologists work. These discussions usually originate in local questions: for example, knowledge debates are entwined with issues of sustainable development in Asia (Agrawal 1995; Bruun and Kalland 1995; Huber and Pederson 1997) and with concerns about predation by pharmaceutical companies in South America (Posey 1990; Brush 1993; Rival 1998). They overlap with land claims struggles in Australia (Povinelli 1993, 1995), in New Zealand (Sissons 1993) and in Canada, where they also concern access to participation in scientific research in the Arctic and sub-Arctic (Ingold 1996; Scott 1996; Nadasdy 1999). Concepts like local knowledge are now in broad circulation and find new points of connection at international conferences, among organisations committed to achieving indigenous rights, and in small communities where access to communications technology is being achieved (see Descola and Palsson 1996; Ellen 1996; Sillitoe 1998; and Ingold 2000 for substantial overviews of this literature).

This chapter provides an opportunity to reflect on how broad public debates about indigenous knowledge or traditional knowledge, being carried out on a global scale, diverge from or shape local debates about knowledge in Arctic and sub-Arctic communities. My impression is that these global and local debates intersect differently in different settings and that to better understand their implications we need ethnographic accounts about how such topics become part of everyday conversation in specific communities. Observations from the Yukon Territory, Canada, may contribute to debates that are now very lively in the circumpolar North.

The paradox that interests me is this. In much of the resource management literature there seems to be a growing consensus that indigenous knowledge exists as a distinct kind of epistemology that can be
systematised and incorporated into Western management regimes. One
of the many lessons we have learned from anthropology is that as soon
as taken-for-granted, everyday knowledge practices become defined
and bounded as 'systems' of knowledge, this can set in motion
processes that fracture and fragment human experience. A recurring
question concerns how knowledge gets identified and authorised in
different contexts, and who gets to control it. More specifically, in
northern Canada, how do different concepts of knowledge connect
with bureaucratic practice in which much of the resource management
literature is centred? I'll begin, then, with a story from one small Yukon
community.

In 1982, the Yukon Historical and Museums Association organised
a conference about early human history in the Yukon Territory. Archaeologists with decades of experience working in northern Canada and
Alaska were invited to participate, along with elders from Yukon First
Nations. The laudable aim of this meeting, unconventional at the time,
was to encourage scientists to meet with local elders in a setting where
they could exchange knowledge about environmental factors affecting
regional history. The conference was hosted in the community of
Haines Junction, headquarters for the Champagne-Aishihik First
Nation, rather than in Whitehorse, the Yukon's capital and a more conventional conference venue.

More than a hundred people – adults and children – crowded into
the community hall on a brisk autumn Saturday morning, many driving considerable distances to be there. Local elders filled the front row
while organisers prepared coffee and assembled slide carousels.
Throughout the day, archaeologists successively presented papers on
past and current research and responded to questions from engaged
audiences. After sitting patiently listening until late afternoon, Mrs
Annie Ned, a Southern Tutchone elder (close to ninety at that time),
rose to her feet asking, 'Where do these people come from? Outside?
You tell different stories from us people. You people talk from paper –
I want to talk from Grandpa.' Thus claiming her authority she began
telling her own stories about subjects of the day's discussions – early
caribou migration routes; trade between coast and interior; her parents' experiences of the Klondike gold rush; her own memories from
early in the century.

When I had originally heard Mrs Ned's narratives several years earlier, I had slim basis for understanding what she was saying. She was

born near the old settlement of Hutshi in the southern Yukon Territory some time during the 1890s before births were formally registered. When we began working together during the 1970s, with the goal of recording her life story, my initial objective was to learn how indigenous women had experienced the tumultuous changes brought to the Yukon during the twentieth century. Rapidly, it became clear that Mrs Ned and her contemporaries were approaching our project with models of life history very different from my own. I had expected our discussions to trace the effects of the Klondike gold rush, missionary-run residential schools, construction of the Alaska Highway, increasing bureaucratic surveillance of women's lives during the 1950s, and so on. Eventually, I came to view that project as flawed by my attempts to impose a conventional academic framework that might evaluate such accounts as historical or scientific evidence. These women kept redirecting our work away from secular history towards stories about how the world began and was transformed to become suitable for human habitation. The more I persisted with my original agenda, the more insistent each was about the direction our work should take. 'Not now', Mrs. Ned would reply to my questions. 'Write down this story about that man who stayed with caribou', or 'Listen to this story about the boy who stayed with fish.' She insisted that such stories were important to record as part of her personal history. If I expected to learn anything, she implied, I needed to become familiar with pivotal narratives that 'everyone knows' about relationships among beings that share responsibility for maintaining social order. While my initial concern was that she was narrowing our focus by insisting on the primacy of traditional stories, it became clear as we continued working together that she was actually enlarging our project. Gradually, I learned how narratives about complex relationships between animals and humans – more properly thought of as human and nonhuman persons – could frame not just larger cosmological issues but also the social practices of women engaged with a rapidly globalising world. Stories *connect* people in such a world, and they unify fragmented or interrupted memories that are part of any complex life. Rooted in ancient narratives, they can be used in strikingly modern ways (Cruikshank et al. 1990; Cruikshank 1998).

On that afternoon in 1982, then, I was intrigued to hear how Mrs Ned would address her audience in the Haines Junction community hall. She spoke at length, relating some stories I had heard before and others new to me, including one about her acquaintance with Skookum Jim, the original discoverer of Klondike gold in 1896 and a subject of earlier discussion that day. 'I almost married Skookum Jim', she commented, startling those of us who knew him only as an emblematic historical figure.

At the time, Mrs Ned's intervention was both singular and memorable for her audience. By the end of the twentieth century, when elders were routinely invited to make presentations to such conferences, this interjection would not be so unusual. There has been a dramatic shift in popular discourse during the intervening years, and the idea that indigenous peoples should represent themselves rather than be represented by others (like archaeologists or anthropologists) now meets widespread, commonsense approval. In Arctic and sub-Arctic Canada, one consequence of this shift has been increasing incorporation of references to local knowledge or indigenous knowledge in public discussion, suggesting that additional voices are being included in public debates. But are they? And if so, how? And if more voices are included, whose remain left out?

In 1992, a decade after Mrs Ned intervened at the Haines Junction conference, I was asked to review a bibliography entitled 'Indigenous Knowledge in Northern Canada', which consisted of 550 carefully annotated entries (L. Howard et al. 1994). It included academic papers and research reports, conference proceedings, speeches, and management plans from all levels of government. But it also included reports by indigenous organisations usually overlooked in academic bibliographies: the Labrador Inuit Association (1977), the Dené Cultural Institute (Barnaby 1987), the Council for Yukon Indians (1991), and the Inuit Circumpolar Conference (1992). The ASTIS-CD ROM database yielded 200 additional references and a sense that indigenous knowledge is indeed a burgeoning field of study in the Arctic.

Moving to international literature, I found materials on indigenous knowledge from Africa, Asia and Latin America. These included reports, conferences, publications and newsletters from UNESCO, the IUCN (International Union for the Conservation of Nature), the World Wildlife Fund, the Bruntland Commission, the World Bank, as well as recommendations arising from the 1992 Earth Summit in Rio de Janeiro, to mention only a few. Overviews include Posey and Overall (1990), Brush (1993), Hansen (1994), Davis and Ebbe (1995), and Kuhn and Duerden (1996). The sheer scope of the literature raises questions about whether there is a growing tendency to present indigenous knowledge as somehow freestanding. If so, does this move us away from questions about what can be learned from local knowledge and, as Paul Richards (1993) suggests, towards assigning reified meanings to abstract concepts?

Critics of this process caution that there is a danger of local knowledge being absorbed uncritically into ideological critique, with opportunities to explore serious alternatives lost in the process. Can terms like 'indigenous' or 'aboriginal' or 'traditional' describe something as precise as *local* knowledge? Such terms are convenient, but conventionally describe a relationship between particular peoples and a surrounding nation-state. As adjectives, indigenous and aboriginal carry symbolic weight yet are broad enough to include a variety of groups with conflicting agendas. Anthropologists have long demonstrated that cultural knowledge is learned and passed on locally. It does not inhere in reified political categories. Yet once the term 'indigenous knowledge' becomes ideologically embedded as 'common sense', it gets welded to other ideas that may inevitably sweep up and submerge knowledge that is learned, shared and passed on locally.

There are larger questions about how we come to frame everyday practices as autonomous, homogeneous, internally bounded objects of knowledge and then set out to analyse them. Much recent scholarship on the history of colonialism traces the specific social, historical and political conditions under which particular kinds of local knowledge were suppressed and eliminated while other kinds, like Western sciences and humanities, gained authority and then claimed legitimacy to analyse forms of knowledge that were marginalised in the process. Local knowledge has been framed repeatedly as a foil for concepts of Western rationality, historically characterised as 'primitive superstition', 'savage nobility', 'empirical practical knowledge', or 'ancestral wisdom'. These terms inevitably reflect more about the history of Western ideology than about ways of apprehending the world. Late twentieth-century recasting of the same ideas as indigenous science or, traditional ecological knowledge (TEK) continues to present local knowledge as an object for science rather than as a kind of knowledge that could inform science.

The burgeoning literature on indigenous knowledge, then, is hardly neutral and has consequences. In northern Canada, three strands especially deserve discussion and in this chapter I simply outline some debates that occurred there during the 1990s. First, there is a growing literature on themes of resource management and sustainable development and, more recently, on the impacts of climate change. Second, much environmentalist writing looks to indigenous peoples for alternative ways of thinking about the environment. Third, indigenous organisations negotiating with different levels of state government in Canada are producing their own materials (including Internet websites) on environmental issues. Although these narratives all address notions of a common good and increasingly share a lexicon, they originate in different domains. The processes by which local concepts become incorporated into Western narratives often seems strangely distant from Mrs Ned's advice about listening to 'different stories'.

Indigenous Knowledge and Resource Management

The language of governance is supple and changing, but it is not always clear how much those changes clarify and how much they obscure. During the early 1970s the Yukon's Member of Parliament, Erik Neilsen, confidently asserted his dream that the Territory's capital, Whitehorse, could become the 'Pittsburgh of the North'. Such views were common then but came under criticism a few years later during the Mackenzie Valley Pipeline Inquiry, when Justice Thomas Berger, appointed to chair the hearings, travelled to small communities throughout the Mackenzie Valley to listen to local views. He concluded that prevailing modes of resource extraction by nonlocal interests should be reframed as a choice between a 'northern frontier' and a 'northern homeland' and recommended against the construction of a gas pipeline through northern Canada until the federal government undertook its legal obligation to settle land claims (Berger 1977).

The explosion of government-sponsored workshops on sustainable development, indigenous knowledge and co-management in the 1990s might be interpreted as a small but significant policy shift. One paradox, though, is the expectation that indigenous northerners should be expected to contribute their observations and interpretations of environmental phenomena only now, in the wake of the Chernobyl explosion, the Exxon Valdez oil spill and recognition that polar latitudes are among the most polluted on the planet. As Feit (1988) suggested some years ago, there seems to be some irony here: the formal expectation developed during the 1990s that northern participants should make these contributions at national and international levels, as members of conference panels and regulatory boards like the Arctic Monitoring and Assessment Program (AMAP) and other working groups and projects under the auspices of the Arctic Council, rather than at a local level where such knowledge might make a concrete difference. Such a formulation seems to suggest that indigenous traditions should provide answers to problems created by modern states in terms convenient for modern states.

Northern hunters have long maintained land-based economies and are undeniably in a position to make unique and finely grained observations. In many instances their knowledge exceeds that of scientists. But that knowledge is encoded both in distinctive paradigms and in culturally specific institutional arrangements for converting those observations into everyday practice, and these may get stripped away in translation exercises because they do not travel easily across cultural boundaries.

The acronym-filled rhetoric of traditional ecological knowledge, or TEK, provides a rich arena for assessing both what happens to local knowledge swept into debates now framed as global, and the consequences when universalising formulations are played back at a community level. An issue of the Canadian Arctic Resources Committee newsletter devoted to this topic begins with an editorial urging that 'an effective system must be developed to collect and classify indigenous knowledge, particularly with respect to northern resources, environment and culture' (Hobson 1992). A newsletter published by UNESCO called *TEK Talk* began its introductory issue with a statement of its purpose as 'further[ing] the recognition and understanding of TEK, promot[ing] the application of TEK in the decision making process and promot[ing] networking among those interested in TEK' (UNESCO 1992). The Commission on Ecology of the International Union for the Conservation of Nature (IUCN) has established a Working Group on Traditional Ecological Knowledge, one of whose objectives is 'to develop and promote ways of harnessing, recording, analysing and applying traditional ecological knowledge for the conservation of nature and natural resources' (cited in Freeman 1992a: 87). And the list goes on.

The overwhelming impression given by such descriptions is that indigenous knowledge is essentially uncomplicated, that acquiring it is primarily a technocratic classification exercise and that managers are those best equipped to identify the appropriate parameters and categories. What seems to be missing in this objectivist paradigm is any sense of how such issues are actually discussed in local communities.

In 1982, one caribou biologist who heard Mrs Ned speak at the Haines Junction meetings later accompanied me to visit her at her cabin, thirty miles north of Whitehorse on the Takhini River. Mrs Ned had referred to the enormous herds of caribou she remembered seeing at Kluane Lake and Aishihik Lake during her childhood. There is archaeological evidence that such herds once travelled much farther south than they do now, though none have been seen in this area in recent memory. Two subspecies of caribou formerly ranged in the southern Yukon – the large herds of Stone Caribou (*Rangifer arcticus stonei*) and the Osborn Caribou (*Rangifer arcticus osborni*), which travel in smaller groups. Scientists are still unsure of reasons for their appearance and disappearance; however, sophisticated analytical techniques enabled by melting snow patches in the southern Yukon during the 1990s confirm that extensive caribou herds inhabited the southern Yukon at least 4,360 BP and probably much earlier (Kuzyk et al. 1999). In 1982, though, this biologist was interested in learning what Mrs Ned could tell him and had specific and careful questions he hoped to ask her.

Of the elders with whom I worked closely, Mrs Ned was the most curious about what scientists and archaeologists were actually up to. She was pleased to see him and provided equally thoughtful answers to his questions. When caribou came in winter, she told him, the sound of their hooves could be heard for miles as they clattered across the ice-covered lakes. On one occasion, large numbers broke through when their weight was more than the ice could support and she described how difficult it was for hunters to retrieve the meat and for women to tan hides soaked and frozen in this way. She went on to tell the biologist about one of the last times caribou came in this direction. A man with shamanic powers disappeared when caribou took him. His kinsmen struggled to entice him back to the human world. They could observe what appeared to be a single caribou on the lake, but once they heard it sing this shaman's song, they understood that he had been transformed and knew what their obligations were. In a powerful voice, she sang for us the song they heard. Gradually and with great difficulty, through a series of elaborate rituals, people were able to bring him back to the human world even though the transition was immensely difficult for him and he was never again able to hunt caribou. She went on to talk about how this story was bound up with her second husband's powers and the story's significance for his life. If Mrs Ned's narrative were to be evaluated by criteria described above it is difficult to say how it would be 'captured', 'classified', 'codified', 'harnessed' or included in a 'database'. In all likelihood, her story would be ignored because it confuses rather than confirms familiar categories.

By the 1990s, Yukon elders were accustomed to holding their own conferences and workshops, and it was managers who were requesting permission to make presentations at elders' conferences. From the perspective of at least some Native northerners, renewed attention to sustainable development can appear increasingly interventionist and intrusive, and consequently these presentations are rarely smooth. At an Elder's Festival in the southern Yukon held in the summer of 1994, a fisheries biologist made a presentation about the contentious catch-and-release programme, which requires any-one who catches a fish below a specified size to release it back into the water. This programme has proven deeply problematic for local elders, who speak forcefully about how such practice violates ethical principles because it involves 'playing with fish' that have willingly offered themselves. The biologist, while expressing sympathy with this position, nevertheless explained as clearly as he could the relationship between fish size and future fish stocks, arguing that rational resource use and long-term management would ultimately enhance the aboriginal fishery. An elder, shifting the field for discussion, responded by reviewing the story familiar to the other participants, of the 'boy who stayed with fish'. A youngster, showing hubris by making thoughtless remarks about fish, trips and falls into a river, where he is swept into a world where all his normal understandings are reversed. In this world, fish occupy the 'human' domain and all the cultural behaviour he has come to take for granted is shown to be foolish and wrong-headed. Gradually, he becomes initiated and properly socialised into his new world and when, the following year, he is able to return to the human world through shamanic intervention, he brings back an understanding of the fundamental relationships enmeshing humans and salmon in shared responsibilities for the health of salmon stocks. Again, it is unclear how this knowledge, widely discussed in the southern Yukon, could be translated into the language of TEK. Biologists working in the Yukon are extremely sensitive to this issue. A booklet integrating various perspectives on this issue was prepared for and distributed by the Yukon Wildlife Management Board (Muckenheim 1998). However, these perspectives inevitably carry less political weight than do quantitative models when political decisions about resource management are reached (Nadasdy 1999).

It does seem that at the heart of Western bureaucratic practice, there is a systematic fracturing and fragmentation of human experience. Taxonomic schemes like TEK tend to work with surface features and are inclined to stagnate and to drain the content – and the life –from their categories. Categorical practices that distance people from lived experience do have consequences. And, as others have argued, these processes are intensified in new cultural orders established as a result of growing bureaucratic, scientific, technical and military management, in conditions sounding rather like those accelerating in the Arctic.

Indigenous Knowledge and the Environmental Movement

In broadly based environmental movements, use of indigenous imagery is more encompassing and historically more complicated than that emerging in TEK. But there is also a more developed critique, much of it coming from anthropology. Technocratic models seem to drain tradition by reducing it to codifiable data. Environmentalist literature more commonly reshapes tradition to fit contemporary concerns. Emerson, Thoreau, Karl May and others independently drew inspiration from what each imagined to be American Indian beliefs and their legacy has continued in the transformation of Black Elk, Chief Seattle and others into environmentalist icons whose words have been reshaped to a present-day idiom. Globalising environmentalist narratives represent themselves as a rejection of modern alienation from nature. Yet they actually reformulate modernism's most enduring narratives – environmental determinism and evolutionary progression – by positing indigenous peoples as part of nature and seizing the mantle of moral truth for environmentalism. This critique does not minimise the seriousness of environmental crises, but rather the tactical appropriation of Native American concepts in these debates. Once again, local knowledge becomes data for Western myth.

The outlines of these debates are well known, and they are based on two problematic axioms: one naturalises Native Americans as 'original ecologists' who are said to have lived in harmony with nature prior to arrival of Europeans. While this position contains elements of truth for particular times and places, it ignores substantial evidence to the contrary (see Krech 1999). It becomes reframed as a backward projection of contemporary views and then goes on to accumulate new baggage. Thus reformulated, Native Americans held long-term perspectives and engaged in highly rational behaviour that sound suspiciously like intuitive foresight of modern management strategies – 'noble savages' with modern Western sensibilities. In Alaska, anthropologists Ann Fienup-Riordan (1990), who has worked with Yup'ik communities for more than twenty years, and Ernest Burch (1994), with more than three decades of research in Alaskan Inupiat communities, have each framed this critique in locally specific ways.

A second axiom presumes that if people in a particular society express respectful attitudes towards environment, they will inevitably behave towards it circumspectly (see Pederson 1995). The argument begins by pointing to Judeo-Christian mastery of nature embedded in models of progress, or Cartesian fracturing of subject and object. Societies that do not so clearly set concepts of nature and culture in opposition, according to this argument, will demonstrate more reverential forms of behaviour towards the natural world. But this view, too, presents a romanticised and overly simplified understanding of the relationships among ideology, norms and behaviour. Sociologists remind us that norms do not determine behaviour in any society, though they may allow us to legitimise our actions in hindsight. In a philosophical framework where animals and humans share common states of being that include family relationships, intelligence, and mutual responsibility for maintenance of a shared world, interaction with the physical world is a social relationship. And, as we know, social relations are rarely straightforward.

The pitfall of both of axioms – one linking hunters with harmony, the other conflating norms with behaviour – is that they each so easily become weapons when indigenous people fail to pass such litmus tests for authenticity, well illustrated in Canada. George Wenzel documents how the animal rights movement first appropriated and incorporated Inuit traditions into anti-harvesting campaigns and ultimately turned those same arguments against Inuit, whom they came to define as enemies because of their use of modern technology for hunting. In the anthropology thus invented by the animal rights movement, Inuit culture was redefined and idealised on Western terms. Hence, 'the word "tradition" becomes a semantic telescope that is used the wrong way round. What is distant is good and what is contemporary is bad because it has been tainted by modernity' (Wenzel 1991: 6; for a Greenlandic perspective see Lynge 1992a).

Despite surface differences, models based on TEK and those based on environmentalism share contradictions. TEK, modelled loosely on ecological science models, heads in bureaucratic directions. Original ecology models, more conventionally phrased in religious terms, posit indigenous peoples as stewards of profound ecological knowledge. Both models force indigenous people to speak in uncharacteristic ways. Both ultimately redefine aboriginal cultures in Western terms by projecting North Atlantic concerns as global.

Indigenous Knowledge and NeoColonial Encounters

These models – TEK and environmentalism – become fraught at their awkward conjunction with the lives of real people. Indigenous leaders now negotiating issues of land and self-governance with state agencies find that the terms of reference for discussion are often already set by globalising debates about sustainable development, environmental

crises, such as pollution and climate change, and indigenous knowledge. In liberal democracies, and more recently in post-Soviet
republics, claims to indigenous status and specialised cultural knowledge currently garner public support, but to do so they must be presented simultaneously as: first, rooted in longstanding tradition;
second, relevant to the modern world; and third, transparently accessible to broad audiences. This is exemplified by the debate over whaling by indigenous peoples. The International Whaling Commission
(IWC) allows Arctic peoples to hunt certain whales under its category
of 'aboriginal subsistence whaling', yet it has been argued that this
implies that indigenous cultures are static and unchanging, and
imposes strict definitional criteria on indigenous peoples – allowing
others to say how whaling peoples such as the Inuit are to be cate-
gorised and also restricting their right to develop markets for whale
products (Nuttall 1998: 119–20). By the late 1990s, too, there was an
emerging populist backlash to acceptance of indigenous knowledge,
fed by economic retrenchment. I want to conclude, then, with two
examples. One concerns the measured success achieved by Canadian
Inuit translating their knowledge as TEK. Another outlines contradictions discerned by a perceptive speaker invited to participate on a First
Nations panel addressing an environmental conference in the western
sub-Arctic.

Canadian Inuit Successes

Canadian Inuit are successfully using models centred on traditional
ecological knowledge, but they may do so at some cost. The devasta-
tion wrought to Arctic economies in recent years by European anti-
sealing campaigns has made Inuit communities extremely wary of
environmentalists – especially Greenpeace – and in many cases they
have preferred to address critical environmental issues by building
closer associations with scientists. They have achieved extraordinary
successes in an international forum. The Inuit Circumpolar Conference (ICC) was established in 1977 and has held official observer status at the United Nations since 1983. The list of ICC's accomplishments
is impressive. They have successfully obtained positions on such international committees as the World Conservation Strategy, the International Working Group for Indigenous Affairs (IWGIA), and Arctic
Council working groups and projects such as the Arctic Monitoring
and Assessment Program (AMAP), the Conservation of Arctic Flora
and Fauna (CAFF), the Arctic Climate Impact Assessment (ACIA), as

well as helping to established the Arctic Council's Indigenous Peoples' Secretariat (IPS). Indeed, ICC has played a major role in reshaping international cooperation on Arctic issues from a concern mainly with environmental protection, under the Arctic Environmental Protection Strategy (AEPS), to a concern with reconciling environmental protection with sustainable development under the Arctic Council (Nuttall 2000). They were prominent at the 1992 United Nations Conference on the Environment in Rio de Janeiro, played a role in the 1993 International Year for the World's Indigenous Peoples, and were active in Arctic Council preparatory discussions for the World Summit on Sustainable Development, held in Johannesburg in summer 2002.

Forging broad alliances can risk disengaging leaders from local issues, but Inuit appear to have successfully linked concerns about global change with local, territorially based knowledge, at least in some communities. Inuit leaders are aware of potential benefits arising from involvement in scientific research, arguably one of the few renewable economic resources in the Arctic, and of the need to secure such benefits for local communities. The Hudson Bay Program, for example, had by 1993 received $1.5 million to document indigenous knowledge (Richardson 1993b). Such projects have helped to define local ideas about how research should be conducted and how knowledge should be transmitted, largely because of their emphasis on local control. Costs may follow, as commoditisation of their indigenous knowledge enmeshes Inuit increasingly in bureaucratic management strategies. The Hudson Bay Program, for example, with its 'Traditional Ecological Knowledge Management Systems' (TEKMS), was described by its director as providing 'a data base for predictive modelling, for forecasting and for selecting harvest areas' and as helping Inuit and Cree 'bring forward their knowledge in such a way that it can be integrated into the cumulative effects assessment' (Okrainetz 1992: 15). If Inuit now gain undeniable advantages by linking their knowledge to the authority of quantifiable and empirical science, claims based on reification of TEK may eventually leave them stranded on a slippery slope between politicisation of that knowledge and its rationalisation by government.

Conference miscommunications

A more circumscribed example comes from my discussions with participants in environmental conferences. Environmental meetings provide a platform where indigenous people may find sympathetic and influential audiences. As with TEK, one option speakers have is to use a rhetoric their audiences expect – even when this forces them to speak in ways not typical in everyday local settings. A more challenging option is to try to complicate perceptions of listeners, but results can be disappointing where audience members bring clear expectations about what they want to hear. A speaker of First Nations ancestry invited to be a panel member at one of these recent conferences reported insightfully about difficulties facing speakers in these settings.

Two elders, and three younger speakers more familiar with such meetings, were invited to address approximately one hundred registered participants. An elder began by telling a story from her childhood about how she and her playmates had once behaved casually and inappropriately towards plants, and how this had resulted in frightening consequences that none of the children ever forgot. She concluded her lengthy narrative by saying that children now are not spanked frequently enough.

My colleague followed. Uncertain about how completely the audience was able to appreciate this narrative and what they might make of the unexpected reference to spanking, she confronted this issue directly. She began by thanking the elder and alerting the audience to the importance of the narrative they had just heard, explaining that it was rarely shared with strangers, and that they were fortunate that an elder was speaking to them so frankly about things usually discussed among closely connected acquaintances. She explained that this was one of many narratives addressing fundamental issues of respect between humans and other living beings. She pointed out that people in her community do not make a distinction between human affairs and the physical environment, and that the Western idea of environment as a redeemable object to be 'saved' was unfamiliar to elders. The reference to spanking, she noted, might best be understood as a concern elders sometimes express about lack of discipline, rather than a representative practice. She went on to talk about how specific stories convey the inseparability of environment from everyday life.

A second speaker adopted a different strategy. He spoke of the concept of respect – respectful behaviour to other human beings and respectful behaviour between humans and nonhumans. Accustomed to using this language to speak to large audiences, he may have seemed to some to overgeneralise the term 'respect'. An audience member requested a clearer definition of the term. The panellist responded, again conventionally, that it is impossible to translate some concepts from indigenous languages to English. My colleague, once again trying to formulate a more optimistic response, proposed that while 'respect' is indeed an English word used only recently, it might be thought of as referring to attention to subtlety, especially in relationships among humans, and between humans and other living things.

A third panellist, astutely observing the difficulty of cross-cultural translation in this setting, framed her presentation with reference to the importance of cardinal points and the sacred medicine wheel to indigenous people. Even though the concept of the medicine wheel lacks historical roots in the sub-Arctic, this panellist could judge that it would be recognisable to her audience and she was the only one of the five to receive resounding applause. In experiments to reach broad audiences who come with universalistic expectations and expect to understand what they hear, familiar strategies are inevitably the most effective in the short term, but they too have costs.

In making public presentations, speakers have limited options. They may construct their arguments by setting local concepts in opposition to Western concepts – partly in order to make the clear distinctions. This usually involves drawing a contrast between an indigenous religious self (grounded in familiar symbols like medicine wheels or respect) and a Western materialist Other. Such boundaries may be rhetorically effective, but asserting them puts indigenous people in the unenviable situation of having to live up to unreasonable and unrealistic standards. In such situations, tradition becomes treated as purity, modernity as corruption and wearing blue jeans or eating pizza as evidence of inauthenticity.

Conclusion

In conclusion, reformulating environmental policies with reference to traditional knowledge may appear inclusive, but, as Phyllis Morrow and Chase Hensel have reported from Alaska, terms like 'co-management' and 'sustainable development' and 'TEK' are highly negotiable and have no analogues in Native American languages (Morrow and Hensel 1992). To suggest that they are somehow *bridging concepts* may constitute taking control of dialogue in ways that mask deep cultural disagreements and restrict the ways of talking about important issues. One of the many things I learned from Mrs Ned and her contemporaries is that their extensive knowledge is not amenable to direct questions, nor can it be formulated as a set of rules. It must be demonstrated so that others can see how it is used in practice. It cannot easily be construed as a written, formally encoded, reified product. Once it is, and once it becomes authorised in this way, it begins to accumulate different meanings.

I conclude with questions that we should ask as anthropologists, rather than with answers. What are the consequences of categorical practices that distance people from lived experience? How does authorising particular kinds of knowledge change its social function? Does the local knowledge of northern peoples maintain its own integrity when it becomes bound into larger narratives? What forms do such transformations take in differing geopolitical circumstances (for example in Siberia)? One contribution anthropology can continue to make is through ethnography that shows how particular local formulations can continue to complicate – and to surprise – universalising, commonsense, expectations about what we mean by knowledge.

Acknowledgements

A longer version of this paper was published as 'Yukon Arcadia: Oral Tradition, Indigenous Knowledge and the Fragmentation of Meaning', in *The Social Life of Stories: Narrative and Knowledge in the Yukon Territory* by J. Cruikshank, Lincoln and London: University of Nebraska Press, 1998, pp. 45–70.

3 Local Knowledge in Greenland: Arctic Perspectives and Contextual Differences

Frank Sejersen

Introduction

As the discussions about the generalities and particularities of local knowledge pervade the academic and indigenous communities, one is increasingly struck by the question: Why has indigenous knowledge never become an issue in Europe and Greenland in the way that it has in North America? (Dybbroe 1999: 14). Although Greenland is an Arctic Inuit community, its history is closely related to Danish colonial policy which sets it distinctly apart from the North American Arctic. An analysis of institutional, cultural and social differences between Greenland and the North American Arctic may provide answers to the above question as well as broaden the discussion of local knowledge in general. The striking differences between Greenland and other Arctic communities with respect to political development, and the role of leaders and the perceptions of community integrity, complicate simple generalisations about local knowledge in the Arctic.

The integration of local knowledge in resource monitoring and management in Greenland has in the late 1990s taken a significant step from simple, political rhetoric to serious dialogue between stakeholders. While the issue of local involvement and responsibility in research and policy making has been discussed in Alaska and Canada for at least three decades, Greenland seems to be on the threshold of this process. It is a striking paradox because one would think that local knowledge should have the best possibilities in Greenland, where an elaborate and extensive political system – when it comes to indigenous self-government – has been developed. The introduction of Home Rule in 1979 has not, as expected, put an end to confrontations and conflicts between stakeholders in management questions. Many hunters feel overlooked and claim that they cannot see much difference between what they experienced during Danish supremacy and what they experience today. It is even been claimed that Nuuk, the capital of Greenland, is as far away as Copenhagen – geographically, culturally and politically (Tobiassen 1998: 207). In some circumstances, the Home Rule structure is perceived as an external, advancing and importunate political system that does not leave much room for local perspectives and inputs. Hunters increasingly find themselves caught in a bureaucratic and centralised web of Home Rule regulations that are difficult to integrate in daily activities and planning. Additionally, hunters complain that important management decisions are made by institutions and staff alienated from the local context. Apart from this kind of institutional criticism, some hunters additionally put forward conceptual criticism. They point out that biologists and their concepts are too dominant in the production of knowledge on resources. Methods to monitor resources are usually perceived to be detached from the local context and local perceptions, as well as carried out by persons detached from the daily activities of hunters and fishermen. Hunters' institutional and conceptual criticisms give them reason to question the research of biologists and the regulations imposed by the Home Rule administration – criticism which has resulted in several conflicts and a widespread feeling among hunters of being marginalised.

Ways of integrating local knowledge and local people into resource management and monitoring is closely related to the way knowledge is framed. In order to understand integration processes and the circumstances under which the knowledge debates gain momentum, it might be fruitful to look at the structure of the political system and the cultural perceptions of community integrity and development. It is a point of departure of this chapter that a focus on the institutional and cultural contexts can be used as a common frame of analysis in comparative knowledge studies.

By examining these two contexts in Greenland, and by comparing them with corresponding contexts in North America, the specific nature of the knowledge discourse in Greenland is rendered more transparent and the diversity in knowledge discourses in the circumpolar North more visible.

Knowledge and the Institutional Context

The introduction of Home Rule in 1979 was indeed a milestone in the relationship between Denmark and Greenland (Gulløv 1979; Foigel 1980; Dahl 1986a, 1986b), but also a milestone in international politics when it comes to indigenous self-government. Over the last twenty years, the Home Rule government has gained more and more political autonomy and is now responsible for most of Greenland's affairs.[1] According to Jens Dahl (1986a: 321) Greenlandic Home Rule can be seen as a 'state' in formation. Home Rule has without doubt given Greenlanders a powerful and flexible tool to deal with their own affairs and future. Although the term Home Rule indicates that Greenlanders are able to rule their own home, the concept of homeland in Greenland has very different connotations compared with the North American Arctic.

The geographical distance between Denmark and Greenland, and the small number of Danes in Greenland, are two of the notable reasons that make decolonisation different from the processes normally encountered in countries with indigenous peoples (Dahl 1993). In modern times, Danes have predominantly looked upon Greenland as the land of Greenlanders (Dahl 1993), while Native peoples in North America have had to negotiate internally as well as externally about the demarcation of areas they considered appropriate as the basis of the land claim agreements. This process involved among other things the demarcation and selection of land, as well as the documentation of historical and present land use and requirements (Freeman 1976; Brice-Bennett et al. 1977; Berger 1985). As a consequence of the complicated land claim puzzles, a very intense sense of homeland, community belonging, and integrity, which should be protected, emerged. Native peoples have been forced to manifest and legitimise their political claims as well as land claims in quite different ways from Greenlanders. In all claims settlements in North America, Native peoples and communities have been required to select bits and pieces of the land they used traditionally. They find themselves caught in a geographical, political and ethnic jigsaw puzzle and the 'Arctic' in North America can be characterised as a number of peripheries within the USA and Canada. In contrast to this situation, Greenland has always been considered a separate entity and one belonging to the Greenlanders. The fact that Home Rule did not have to be worked out in relation to land use and occupancy generates totally different preconditions for nation building, and frames the knowledge discourse in other ways.

This important difference has consequences for the ways institutional settings are constructed and perceived. In North America the state's wildlife management systems are pervading many local hunting activities and are often considered intrusive. From a Native point of view, management conflicts can be perceived as being based on a clash between outsider vs. insider; colonial vs. indigenous (Usher 1986). Consequently, the state wildlife management systems have been criticised by Native peoples because they believe they erode Native peoples' integrity and community autonomy. The state system is, according to Gail Osherenko (1988), fundamentally ill-suited to Native communities and compliance with governmental regulations is low. A variety of co-management regimes (Osherenko 1988; Huntington 1989, 1992; Adams, Frost et al. 1993; Richard and Pike 1993; Collings 1997a) has been established in order to make the management regime more compatible with local/indigenous management institutions, knowledge and needs. Indeed co-management regimes have proven to strengthen the integrity of native peoples, to a great extent.

The subtle difference between the North American Arctic and Greenland lies in the ways colonial histories have unfolded, how development strategies have taken place and how they are perceived. In Canada and Alaska the colonial structure and politics are perceived by Native peoples as an invasion that has destroyed Native management institutions and practices as well as the possibility of practising one's rights. Loss as a result of the process of destruction can be pointed out as a major ingredient of the development discourse. In Greenland, people relate to the Danish-dominated development as alienated and alienating from the Greenlandic context.[2] Hence, alienation is the major ingredient of the development discourse in Greenland (see for example, criticism of the Danish development policy by Lidegaard 1969 and Olsen 1969). Loss and alienation seem to be two closely related concepts – but they imply different ways of relating to knowledge, institutions and culture, as will be indicated throughout this chapter.

Managing Relations between Local and Central Institutions

The first important political institutions in Greenland to deal with Greenlandic matters were the two provincial councils which functioned between 1908 and 1979 (they were merged into one council in 1950).[3] The councils discussed Danish political initiatives and put forward suggestions and issued regulations themselves. These councils gave Greenlanders an opportunity to influence Danish policy making to some extent, because the councils were consulted and gave their opinion on most issues.

The provincial councils were the first, major Greenlandic institution to influence hunting and fishing. However, the local regulations which existed prior to, and coincidental with, the regulations imposed by the councils have seldom been pointed out as Native/ indigenous/ traditional management systems by proponents of the present management discourse. The only management system that has been labelled as traditional was the presence of traditional hunting territories (Petersen 1963, 1965; Brøsted 1986; Haller 1986; Dahl 1998, 2000). Traditional hunting territories were gradually considered politically unsuitable for modern Greenlandic conditions both by the provincial councils (see for example, Grønlands Landsråds Forhandlinger 1961: 68) and later by the Home Rule government. The diminishing importance of hunting territories has resulted in few conflicts (for exceptions see Forchhammer 1992; Dahl 1998, 2000), as most hunters have increasingly started to use an extensive area for hunting and fishing. Hunting territories are primarily considered a hindrance for free mobility within the Greenlandic sphere. The breakdown of hunting territories was a result of the growing emphasis placed on Greenland as one territorial and administrative entity within which all residents should have equal opportunities and access rights to resources. The continuous process of nation building has fostered political centralisation. Nicolai Heinrich (Anonymous 1998a), a hunter from Nuuk, recalls the relationship between the councils and the Danish political system and considers the previous political setting more receptive to local concerns and inputs compared to the Home Rule setting today. The councils can be perceived as institutions which, at that time, were considered to guarantee the integration of local perspectives and interests into the administration of Greenland.[4]

In Greenland today there is little talk of co-management or decentralisation because the Home Rule government, municipalities and interest organisations are considered the principal actors in management affairs and negotiations. At the moment, management is mainly centred around the Home Rule administration, which is the principal formulator of hunting and fishing regulations, as most issues are considered national. Although the authority of Home Rule is widely accepted, this structure occasionally fosters local dissatisfaction with the central administration in Nuuk, which is accused of being Danish and insensitive towards local needs. Paradoxically, the Home Rule structure reproduces a similar kind of alienation from local affairs and conditions for which Greenlanders accused the Danish system. Part of this may be explained by the nation building strategy pursued by the Home Rule government. Greenlandic Home Rule, to quote Jens Dahl, 'was a political reform, recognising a politically, geographically and demographically undivided Greenland' (Dahl 1986a: 323). This nation building process, with its centralisation, unification and omnipotent perspective, has been further strengthened by a new election act that abolishes the former system of eight constituencies. Now, Greenland has been turned into one single constituency, and representation by marginalised regions (East Greenland, Avanersuaq, Uummannaq and Upernavik) is not guaranteed in the parliament any longer.

Although the clashes between local and central perspectives are understood in ethnic terms, they are not similar to the ethnic clashes encountered in the North American Arctic. Here, there seems to be a sense of two distinct ethnic and political systems – an inside and an outside – which constantly clash to the disadvantage of Native peoples, who have lost sovereignty. To bridge this gap local and regional management bodies have been set up to interact with state institutions.[5] The decentralised system encountered in the North American Arctic is built on an acceptance of Native users as meaningful actors in monitoring and management of resources. The possibilities for Native people are, among other things, the outcome of the Native claims agreements. These agreements have laid down a fundamental political basis in the North American Arctic that supports different degrees of decentralisation and constitutes possibilities for Native action and involvement. However, this may not necessarily change the basic feeling of some First Nations that they are only clients in a state management regime.

The ongoing struggle of indigenous peoples to end colonial hegemony in the Arctic and thus to regain control and recognition has produced different political settings in each of the Arctic states, which have different implications when it comes to the mapping out of de jure and de facto rights and possibilities in resource management and monitoring. Greenlanders manage their extensive rights to self-determination under Home Rule in a very centralised fashion, giving little space for regional input and responsibilities, while Native peoples in Canada and Alaska manage their restricted rights in a decentralised fashion, giving more space for different regional (and ethnic) input. The circumstances for the integration of local knowledge and people are thus quite different.

Local Knowledge in Resource Monitoring and Policy Making

In Greenland, local perspectives are primarily voiced through national associations and political institutions. Management negotiations take place between rather few stakeholders (biologists, the Association of Municipalities in Greenland (KANUKOKA), KNAPK (the National Association of Fishermen and Hunters in Greenland), and Home Rule Departments). The Home Rule Department responsible for hunting takes up an important function as a buffer, and tries to integrate as many perspectives as possible, including its own. However, the limited number of negotiating partners and the centralised political structure in Greenland often make hunters feel marginalised. While this feeling of marginalisation may indeed be valid, managers and biologists often state that the hunters' interests are so dominant in the elaboration of management regulations in Greenland that they, as political and scientific advisers, sometimes feel totally ignored. This is primarily due to the powerful Association of Fishermen and Hunters in Greenland (KNAPK), and the general political goodwill towards hunters. Indeed, it is difficult to draw a one-sided picture of the structure of power, control and influence in Greenland.

In the case of hunting and fishing, the Association of Fishermen and Hunters in Greenland is the most prominent. Since its establishment in 1953, it has successfully lobbied to implement regulations that have created good opportunities for hunters and fishermen. Today, there are seventy-two local organisations under this umbrella association. KNAPK's lobby work can be very effective politically but may not necessarily leave much room for particular community or occupational strategies. The consultation and negotiation process is based on a hierarchical system of representatives and experts. KNAPK staff are consulted as representatives – biologists as experts. Local hunters are not always very visible in the system as such. This is a paradox, considering the diversity of needs, perspectives and problems of the members of KNAPK. Consequently, several groups of hunters and fishermen have criticised KNAPK and have created their own interest organisation (see Sejersen 1998: 255ff.). However, the system is slowly changing and has started to include a diversity of local voices. Decentralisation and integration of local knowledge were, for example, discussed by the Greenlandic parliament in spring 1999. Biologists are also looking for new methods to integrate local knowledge and have taken initiatives to establish platforms for the exchange of knowledge on several species between biologists and local hunters.

This openness to both diversity and decentralisation has been interpreted by some as a threat to Home Rule sovereignty and the scientific integrity of biologists. The latter concern has been advanced by, for example, senior research biologist Mads Peter Heide-Jørgensen. He is of the opinion that the quality of scientific work has to be evaluated by independent researchers from the scientific community itself (Heide-Jørgensen 1998: 134) – not by local people. The integrity, quality and independence of research are maintained when biologists are freed from the hunters, he argues. 'Biologists should not be forced to use hunters' knowledge; [biologists] can be forced to listen to it, but not to use it. Biologists ought to be free and independent. Society receives the best investigations this way' (ibid.: 134).[6] His statement raises questions about 'open-ended' consultation processes and participatory approaches – in other words, exactly how far should local people be consulted or integrated as project participants? A number of biological research projects in Greenland, including one by Mads Peter Heide-Jørgensen, integrate local hunters and their knowledge in order to improve the scientific results. Thus, local knowledge is being used to some extent, but it is firmly located in the hands of scientists in order not to weaken the scientific integrity of researchers and the research results. Rather than seeing such a step as a positive way to improve science, Paul Nadasdy (1999), on the basis of fieldwork in Canada, criticises this way of dealing with local knowledge because it concentrates power in the hands of scientists, who grant themselves the position to set the standards of relevance. New ways of combining the community level and the central administration/research institutions in Greenland are, on the one hand, an attack on the existing authorities but, on the other hand, they may also further a democratic process and new communication routes which can establish new strategies for sustainability based on the building of new relationships between people. For such a process to be successful, the integrity of all stakeholders must be considered, not only the integrity of local people.

Despite worries, the call for the integration of local users in resource monitoring and policy making has been taken up positively by biologists and their research institutions (Anonymous 1999; Pitu 1999). This call also gives rise to the experimentation with the establishment of meeting grounds where hunters/fishermen and biologists can meet and exchange views and information (see for example, Anonymous 1999). Some conceptual obstacles exist, however. An example is the scientific monitoring of caribou in the selected index area in West Greenland, which at the beginning of the 1990s made biologists conclude that the stock was under pressure. As a consequence of the estimate, politicians reacted with a two-year hunting ban from 1993 to 1995. Hunters were outraged and convinced that the biologists were wrong. They felt that their experience was overruled and overlooked by biologists and politicians. The political aspect of the conflict fuelled the hunters' discontent with biologists and their counting methods. Caribou hunters found a few days of counting from an aeroplane inadequate because this method does not really consider the complex behaviour and migration pattern of caribou as observed by hunters. Hunters try to create a complex picture of caribou behaviour, which they construct on the basis of a multitude of their own and other hunters' observations and interpretations, and they consider it as more correct than the scant picture constructed by biologists on the basis of – what the hunters perceive to be – a few days of work. That biologists in fact work with a multitude of factors does not dispel the hunters' basic feeling that biologists really do not know what they are talking about because biologists are outside the continuous discussion of the whereabouts of caribou. This perception of biologists' social distance is intensified by the physical distance that, for example, becomes manifest when biologists use aeroplanes.

When estimating the state of affairs with respect to caribou, both hunters and biologists integrate a number of factors like caribou presence, feeding potentials, calving success, migration pattern, natural death, human activity etc. Hunters desire to encounter caribou of a certain quality and they may decide to increase their activities in a certain area at a certain time, depending on a number of factors, in order to increase their hunting success. Biologists, on the other hand, focus on the caribou stock – a demarcated, quantitative and abstract entity (Roepstorff 2000: 171, 174). In order to see this 'abstract caribou group' and its fluctuations, they are preoccupied with ensuring that the counting methods do not undermine survey viability and comparability. Thus one reason for the polemic between hunters and biologists is the different parameters used to evaluate success. For hunters actual seeing and shooting caribou are success parameters. For biologists comparability and viability of survey methodology are success parameters. Seeing caribou is not the primary goal of biologists. A survey can be a success without even a single animal being spotted. The different emphasis and interpretation of what seeing implies has led hunters to criticise the biologists for choosing the wrong locations or wrong methods, which could diminish the chances of seeing caribou. Hunters are used for counting when doing aerial surveys and they speak from experience when they criticise the method. In their view, the speed of the plane makes the landscape shimmer and consequently it is difficult to see caribou.

Hunters very often argue that the number of caribou in the landscape primarily depends on natural cycles and natural reasons. Thus, hunting bans will not counteract a decrease in caribou numbers to any great extent. Biologists acknowledge natural cycles but perceive hunting to be a dangerous addition to a naturally caused decrease. The role of humans is indeed at the core of the problem. In a series of newspaper articles (Anonymous 1998b; Stubkjær 1998), a hunter puts forward his view of the matter; it represents an interesting case as the subject is locally discussed. He is convinced that the warble fly (*Oedemagena tarandi*) is the reason for the decreasing number of caribou, and not hunting. According to him the fly attacks single and young animals and its parasite causes death to the host animal. He concludes that caribou management should integrate this aspect and create a management system that stimulates the shooting of exposed individual animals. A caribou biologist responds in an interview (Anonymous 1998c) that a thousand caribou dead of natural causes (for example from warble flies) is what can be expected on the basis of the size of the present stock. However, the biologist continues, a recognised survey in Norway indicates that only one out of 1,300 reindeer in Norway was not attacked by the parasite. The biologist concludes that the parasite does not want to kill its host and the death of reindeer is caused by several combined factors. In an aggressive and condescending response (Stubkjær 1998), the hunter attacks the use of Norwegian studies of reindeer to understand Greenlandic caribou, which have only recently (1952) been exposed to the warble fly.

The controversy reflects different points of departure. The hunter opposes the biologist's understanding of Greenlandic caribou on the basis of mathematical stock estimations and 'imported' surveys from outside Greenland. The hunter accuses the biologist of being detached from the Greenlandic and local reality (for example, he urges the biologist to 'move your ass, turn off the computer, and get out in nature' (Stubkjær 1998)). The responses from the biologist reflect his view that even local incidents of caribou death should not come as a surprise due to the general size of the stock. The local perspective (and worries about the observation of several mysterious caribou deaths) is overruled by a generalised perspective associated with the concept of a stock.

In order to bridge the gap between people with different knowledge traditions,[7] Sillitoe (1998) suggests that interfaces should be established. Speculations on caribou behaviour and external factors could constitute an interface between hunters and biologists – an interface both groups would benefit from. With regard to the difference between biologists and fishermen in the perceptions of the halibut's life cycle, Andreas Roepstorff (2000: 183) suggests that a focus on the organism in its environment constitutes such an interface. 'It would...be wrong to see...two mutually incompatible "systems of knowledge" opposing each other.' He considers this meeting point/interface an equal one because it is not alien to either of the groups:

It would seem that both biologists and fishers could benefit from a common 'organisms in their environment' framework. First of all, both groups would be able to contribute information that could be useful and interesting to the other party. Biologists would be able to access much of the context-bound, specific, local information that fishers possess but biologists do not because they cannot be there all the time. This framework might also help ground scientific concepts, which are necessarily general and global, in local contexts (ibid.: 185–6).

Similar to the halibut case described by Roepstorff, the interface for caribou hunters and caribou scientists could be the interrelationship between caribou and their surroundings and thus it could redirect the discussion away from stock estimations – an area which is indeed characterised by disagreement and conceptual monopolisation by biologists. In 1999, the Greenland Institute of Natural Resources entered an agreement with the hunters' organisation which improves the integration of local knowledge and future cooperation.

The hunters' accentuation of biologists' or managers' lack of practical and daily experience as well as the structural distance, socially and physically, can be found in many management conflicts (see for example, Vestergaard 1992; Pálsson 1994). 'The perception of structural distance is like provincialism in Denmark, in the sense that it expresses peripheral discontent with central authorities' (Petersen 1991: 21). However, this structural distinction felt in Greenland contains an additional element: it is perceived to be rooted exactly in the kind of alienation which one encountered during Danish dominance. Despite the presence of a Greenlandic self-government, many conflicts are still understood in colonial terms. The hunters criticise biologists and resource managers for thinking in 'Danish' ways (read 'alienated'), either because they are Danish, or because they are Greenlanders educated in Denmark. Even though one cannot say that the hunters are indigenous people oppressed by a colonial state, they sometimes present their relationship to the Home Rule administration as a colonial relationship (see e.g. Petersen 1994). When the Greenlandic power hierarchy and diverse ways of perceiving things are understood to be of a colonial nature (see also Petersen 1995), management discussions become very complicated and sometimes fuzzy owing to the dichotomy Danish/Greenlandic (dominating/dominated). One can get stuck with the simple colonial dichotomy that offers an easy frame of reference when disagreements are to be understood. This situation is not unique to Greenland. Discourses on development and resource management in many places in the world are contextualised with references to 'locals' vs. 'newcomers', 'resource users' vs. 'planners'. In

Scotland, for example, rural development issues are often phrased as a conflict between 'locals' and 'incomers', between 'black natives' and 'white settlers' (Jedrej and Nuttall 1996). In its simplest form in Greenland, disagreements are thought to be solved by urging the 'Danish' opponent to return to Denmark (as advocated by Petersen 1994).[8] The Greenlandic professor Robert Petersen explains the problem in the following way: 'We are so accustomed to interpreting problems in the context of colonisation that, even today, we try to interpret common political disagreements as a kind of internal colonialism, e.g. as remote government either from Copenhagen or from Nuuk. This is probably also an effect of the colonial past' (Petersen 1995: 125).

This conceptual starting point conflicts with the institutional setting and the possible solutions. It is difficult to find an institutional interface that is based upon and can address this perception. In Greenland there are few institutional platforms or interfaces where people have the opportunity to meet and discuss relevant management issues or exchange knowledge. When considering solutions and interfaces it is also of paramount importance to examine cultural perceptions of what constitutes important knowledge and where it can be found.

Knowledge – Perceptions of Authority and Community Integrity

Generally speaking, people ascribe value and importance to specific kinds of knowledge, and its position in social and cultural changes. This affects how people relate to individuals producing the knowledge singled out as prominent. Normally, scientists' knowledge has high status in many societies, while local people's knowledge has low status. Specific groups and/or people are acknowledged as the producers/bearers of knowledge, and they may get a prominent say in the direction of change and in the efforts to achieve what is perceived as the good life. In the following, the identification and status of knowledge producers in the North American Arctic will be compared to Greenland. The comparison will indicate the very different contexts in which the transmission of knowledge takes place.

Keepers of Knowledge in the North American Arctic

In the North American Arctic elders take a prominent and central position in the transmission of knowledge. Elders in Alaska, for example, are recognised in the sense that they are nominated for their knowledge of land, resources and native culture. In 1996, to give an example, elder Paul John from Nelson Island received state-wide recognition in Alaska for his leadership and continued efforts to ensure that pieces of the Yup'ik past are carried into the future, when the Alaska Federation of Natives named him Tradition Bearer. Elders often point out that some of the young Native people, who could benefit from the knowledge shared by elders, are engaged in other activities and go to school where the teaching may be radically different in form and content (see for example, the statement in McNabb 1991: 65). This gap between generations has been the issue of many workshops, symposiums and conferences in the North American Arctic. Without the knowledge possessed by the elders it is believed – especially by elders – that the young generation will move away from what they perceive as the path towards the good life. According to Kenneth Peter of Akiachak in Alaska, 'Your ancestors made a path for you! You will not go by any other path!' (Yupiit Nation 1991 quoted from Fienup-Riordan 1997: 110). This sense of deviation from the right path was also touched upon by Larry Merculieff at the workshop on Alaska Native Traditional Knowledge and Ways of Knowing, convened in 1994. He is quoted as saying that Alaska Native young people:

> believe that they do not need to listen to the old people. The young people are not paying attention to their own culture. They are asleep within their own culture and are referred to as 'dreamers'. It is time to provide a wake up call to the young people. It is time to re-establish the rightful position of the leaders. It is time to refocus on Alaska native cultures and the wisdom they hold. (Fehr and Hurst 1997: 6–7).[9]

In order to pass on the wisdom[10] of elders, the workshop suggested that groups around the state should list their basic and traditional beliefs and have them posted up everywhere around the communities as constant reminders for behaviour (ibid.: 9). The traditional world-view is considered to be in total contrast to Western cultural values, and thus constitutes the key to change the misdirections of today (see also McNabb 1991: 65ff.). Merculieff creates a dichotomy[11] where Western culture is presented as going to extremes, void of spirit and where people talk a great deal, whereas Native culture lives in balance, is filled with spirit and listens. According to Merculieff, this lack of respect from Western people is repeated when Western scientists come into communities without getting permission or consulting elders. The workshop appointed the elders as the ones to be contacted first in a community and asked for consent and approval before any traditional knowledge is shared. It is their wisdom that will provide direction for the utilisation and stewardship of Alaska Native understanding of the world (Fehr and Hurst 1997: 14).

Despite the prominent position of elders as bearers of knowledge that is perceived as important for community and cultural survival, the position of elders is challenged by some of the young native people. Charles F. Hunt, an Alaskan Yup'ik, believes that his elders are in denial concerning the enormous social and economic changes that have occurred in their homeland over the last fifty years and that they have a difficult time understanding what is going on or how to solve the problems. However, he points out that elders still play a prominent leadership role and many young people fear opposing the elders (Fienup-Riordan 1999: 16–17).

In Alaska, Canada and Greenland, elected representatives – be they corporate, communal, regional or national – play an increasing role in determining the direction of change. In this process, elders in Alaska, but also to a certain extent in Canada, feel overruled. In a poem, Gary Raven (1996: 52–4), an elder from Hollow Water First Nation, Manitoba, looks critically at some of these modern processes of choosing representatives, which have threatened the traditional power hierarchy:

> Today
> New ideas come into the community
> Self government
> This has created a lot of problems
> People in power are benefiting
> Elders lose control
> If we neglect our people they will end up in jail

These worries about elders losing community control and opportunities to guide the youth echoes the worries expressed in Alaska. Furthermore, Raven criticises the new political leaders for being detached/alienated from community life (ibid.). Like Merculieff, Raven sees elders as the bearers of knowledge. Elders possess important knowledge for the survival of the community and a harmonious way of life (see also McNabb 1991: 63f.). Today, consultation with elders takes place at healing circles, among other places. At these circles, guidance by elders and traditional knowledge are part of the healing process. In North America, the past and the traditions seem to constitute very important points of reference for the good and harmonious life, and elders occupy a position as living links to this past.

Knowledge Producers in Greenland

In Greenland, the role of tradition and heritage in development and modern times has not been very salient. Greenlanders have had and perhaps still have an ambiguous understanding of their past: 'Vocally articulate Greenlanders demanded progress and independence for their people, though at the same time demonstrating a high degree of ambiguity and insecurity towards their aboriginal past' (Dybbroe 1996: 43). On the one hand, they are proud of their past and ancestors (see for example, interview in Nilsson 1984: 86), and, on the other hand, they maintain a distance to their forefathers, who are considered heathens and primitives. They make a cleavage between past and present – between the Eskimo way of life and the Greenlandic way of life.[12] Louis-Jacques Dorais (1996: 29) explains this cleavage as a direct effect of acculturation and thus loss of culture. While Danish colonisation and presence in Greenland, since 1721, can explain such a cleavage which is not found in other Arctic communities (according to Dorais), another and perhaps more important question should be asked: Why does this supposed cleavage between past and present not play a significant role in Greenland? The past seldom constitutes the image of the good life and is thus not used as a point of reference to the same degree as in the North American Arctic. Today, the good life – not understood in any absolute sense but as an image comprising several layers of meaning – is thought to be obtainable when Greenlanders get the necessary political and administrative position. The process of managing self-determination and control rather than in reviving or reinstating specific cultural elements from the past defined by the elders, seems to be the strategy in Greenland (see also Tobiassen 1998). Cultural strength is pursued by creating good living conditions in a stable, modern, strong and cohesive nation – Kalaallit Nunaat, Greenland.

Old People in Greenland

Following the, to some degree, insignificant position of the past, elders occupy a totally different position in society today. It is reasonable to say that there are no elders in Greenland, only old people. Obviously, they may play an important role in the private sphere, where they are treated with respect. On some occasions in some families, the words of old people carry weight. But they do not occupy the acknowledged public position of authority as in Canada and Alaska, where they, for example, are appointed as the ones to consult and contact to get approval or knowledge. Some even express the opinion that elders in Greenland are too reserved and too submissive (Nilsson 1984: 48, 103).

In a Home Rule report of the Culture Committee (Kulturrådet 1990: 44f., 62), old people are mentioned very briefly, and the committee suggests that they should have the opportunity to meet with each other and to be consulted as a way of passing on the cultural heritage. The report urges people not to 'put away'[13] the old people and look at them as old–fashioned and against progress. The committee would like to see old people given more respect as in the old days (ibid.: 44). At a recent gathering of old people in Greenland, in the town of Sisimiut, the issues discussed primarily dealt with how to keep in physical shape, and how to organise one's life to maintain a good and healthy attitude and social life in old age. The transfer of knowledge from the old to the new generation was only touched upon very briefly (see articles on the old people's gathering in the local newspaper, *Sivdleq* no. 22, 1999).

Today young Greenlanders make their own path – to use the Alaskan metaphor (see Fienup-Riordan 1997) – and by doing so create new possibilities for the good Greenlandic life with a focus on political control, economic development, employment, better living standards and education. Specific cultural values and elements such as Greenlandic language, Greenlandic foods, village and family solidarity as well as outdoor activities (often condensed in the community life of small villages)[14] are identified as important ingredients in a Greenlandic way of life and necessary to maintain a healthy Greenlandic identity (see Sørensen 1994; Sejersen 1998; as well as the interviews in the book by Nilsson 1984) – but these aspects are not necessarily associated with old people, the sharing of traditional knowledge or a canonisation of traditional cultural values. The good life is perceived to be possible by gaining control over development and by giving it a Greenlandic dimension by aspiring to employ Greenlandic-speaking Greenlanders as workers, administrators and politicians. No Greenlandic values are defined and explicitly presented as important in the development process (Dybbroe 1989: 152), contrary to what we see, for example, in Alaska (McNabb 1991: 65). As long as the Greenlanders themselves are in charge, it is perceived as development on Greenlandic terms. This is, among other places, reflected in the Act of Home Rule (Hjemmestyreloven), where the vision for a more Greenlandic Greenland is limited to five dimensions (see Tobiassen 1998: 166). Only one of these deals with culture: language and the preservation of villages (*bygder*). The latter dimension, which aimed to uphold a decentralised settlement pattern, was among other things a response to the social problems encountered in the wake of the Danish population concentration policy from the 1950s to the 1970s. In order to implement this dimension, a Department for Small Villages (Bygdedirektoratet) was established in 1979 to ensure the development of social and economic welfare in these villages. Despite some good but uncoordinated results, the Department was closed in 1990, and its functions taken over by other departments. An analysis of the Home Rule policy to preserve the small and scattered communities concludes that it was symbolic and sentimental as it lacked concrete vision and economic priority (Lynge 1991). Today, some Greenlandic politicians are in favour of centralisation and the abandonment of uneconomic small villages.

Greenlandic Leaders

Historically, authority in Greenland was primarily restricted to the household level but was occasionally expanded to include temporary leadership of a group pursuing a specific task within a limited period (Nooter 1976; Sonne 1982; Petersen 1993, 1998). No one had authority over the community as a whole. This leadership structure changed when the Danish administration and church introduced positions that had authority in community affairs. These authorities were respected and feared. An individual considered to be asserting personal superiority is described as *naalagaaniartoq* ('someone trying to be better than others') and such a person often 'demands obedience' (*naalaqqutooq*) (Nuttall 1992: 162). There was no resemblance between the colonial hierarchy of authorities and the traditional system of authorities (for a discussion of differences see Nooter 1976). Locally, the great hunters (*piniartorsuit*) who distinguished themselves on the basis of their achievements were highly respected also outside the household, as their obligations were centrifugal. Even today, hunters also have a quite clear picture of who is a good hunter and who is not. But they are seldom directly identified or called great hunters and the hunters rarely single out themselves as more experienced or successful than others – they do not wish to claim that they are superior. When asking about whom biologists should contact, one can encounter answers from hunters like 'our organisation', or 'there are a lot of hunters who have now become very experienced'. In some cases, however, community members may point out a specific hunter because of his knowledge and experience within a certain and limited field. In this sense leaders or experienced hunters who are community or hunter representatives cannot be clearly identified, although they may actually play an important role in social affairs (Nooter 1976; Dahl 1989; Petersen 1993).[15]

Today, leaders and representatives are needed in order for the hunters to deal successfully with hunting regulations, political meetings, and anthropologists (and now biologists) who wish to visit their communities. The local organisation of fishermen and hunters (KNAPK), present in most communities, has an important role as an official and accepted mediator and gatekeeper. Specific hunters in the community do not have by definition a seat on the board because of their experience or age. Sometimes well-respected hunters are chosen as chairperson, sometimes not. Owing to the workload as board member, the position is not necessarily attractive.

In a way this institutional setting mediates between a highly hierarchical structure in Greenland with recognised representatives and negotiators, on the one hand, and a local social setting with no official leaders who can represent the community as such, on the other hand. The local organisations play a significant role as formalised institutions that can open the door to the local hunter community and legitimately suggest hunters to be consulted. But the question of representation remains unclear owing to the diversity of hunters within the bigger communities. This differs significantly from the way communities are thought to constitute a more integrated whole in the North American Arctic, as will be discussed below.

Perceptions of Community Homogeneity

In Canada, land claims negotiations, community hearings, and land use and occupancy projects (e.g. Freeman 1976) have created a political environment centred around traditional territories and Native communities, on the one hand, and federal/provincial land and government, on the other. Apart from the many organisations created in the wake of Native claims and implementation processes, local communities have gained experience in fields of politics, representation and community territorialisation. These claims have been strong statements about boundaries and belonging. The political process has indeed been a struggle to gain control over demarcated geographical and social space, and it has resulted in recognised hearing, consultation and co-management arrangements, which empower First Nations and communities in general. Empowerment, and thus the wish to control as many aspects of relations affecting the communities as possible, has among other things resulted in a new focus on research in Native communities and how its topics are defined, pursued and analysed. Inuit are insisting on being treated as equals in the northern research agenda (Flaherty 1995). One way to empower Native people and communities in this respect has been to claim that '[t]he community (collectively) owns the information shared by individuals. They have the right to control what information is removed from the community and published' (Oakes and Riewe 1996: 76). When community borders are drawn to overlap the political control to such a degree that individual statements/information are owned by the community, the community appears as a homogeneous, undivided and uncontested whole with a total system of values.[16] Martha Flaherty, Pauktuutit President, makes it clear in a firm way that partnership between researcher and Native people is not enough. 'The knowledge you [researchers] gain comes from our ideas, beliefs and traditions. Our right to ownership of the direction and findings of research cannot be contested or denied' (cited from George 1996: 15; see also Flaherty 1995). The strong sense and claim of community homogeneity and control fails to reflect upon the question of representation: who defines how the communities control and own what kind of information and knowledge to whose benefit? Anthropologist Louis-Jacques Dorais, from Canada, fears that total community control over research could have a negative effect: 'They had it in the Soviet Union and China for many years and there was complete control and we saw what it did' (cited from George 1996: 15). According to Ann Fienup-Riordan (1992), Alaskan Yupiit leaders downplay conflicts and disagreements in order to 'work with one mind' and to be successful in their lobby work. Yup'ik ideology, as it relates to political activity, has two fundamental features that help to explain their ability to work for unity: (1) people's duty to pursue the path they are taught; and (2) their need to work with one mind (ibid.: 79–80). These two features make it possible to unite while respecting different community paths. Respect for unity and diversity are, according to Fienup-Riordan, an essential element of Yup'ik political ideology. The promotion of images of communities as homogenous entities indeed has political potential (Li 1996; Nuttall 1998: 166ff.).

The sense of community belonging is quite strong in Greenland (Nuttall 1992), but one does not encounter the same need to promote an image of communities as undivided wholes striving to gain political control. Much of the political discourse in Greenland is about how the elected representatives of the people manage, or rather mismanage, their positions – and how an elected person belongs to a specific faction within a diverse community. There is not the same sense of urgency about protecting community borders from outside intruders (biologists may constitute an exception). This is reflected in the virtually nonexistent discussion on social sciences research and research ethics in Greenland.

Indigenous peoples in the Arctic are aiming to make changes at all levels of society and to regain self-determination. The direction of change and tools suggested to implement the desired change are different in Greenland from those encountered elsewhere in the Arctic. Three interrelated aspects crystallise from the above discussion: first, no publicly acknowledged authorities and leaders (such as elders) are thought to possess the knowledge for correct development; second, a modernisation strategy on Greenlandic terms is pursued rather than a strategy based on a revival of cultural values stemming from a perceived traditional past; third, the fear of a gradual erosion of community integrity and control does not play a significant role when knowledge is exchanged with 'outsiders'.

Conclusion

The tremendous amount of scientific and political attention that Indigenous/Traditional Knowledge has gained worldwide must not lead to simple generalisations. Even among Inuit, the knowledge discourse takes various routes and is framed differently. Indigenous peoples in Canada and Alaska have achieved their aim of putting Traditional Ecological Knowledge (TEK) at the top of numerous agendas. In North America traditional knowledge has become politicised to a great extent, while Greenlanders seem to relate to and talk about knowledge in a different way. The knowledge debate in Greenland has taken a different turn because it is framed and grounded in a completely different political, social and cultural environment from the one encountered in the North American Arctic. Apart from diverse colonial histories and modes of historical representation, the knowledge discourse is also influenced by diverse ways of accentuating and defending community borders as well as dissimilar ways of perceiving authority (leaders, elders, great hunters and political representatives). Finally, it is argued that perceptions of development directions and ways of pursuing the good life frame the discussion on knowledge, too.

Politicisation, exchange, representation and promotion of knowledge are activities that are deeply embedded in very diverse cultural, social and political settings. Basically, the Arctic indigenous peoples are situated in different contexts where they have to manage the relationship between local and central power in different ways. One example is the disagreements between local hunters and centrally based biologists and managers about stock estimations and strategies for sustainability throughout the Arctic. In the North American Arctic, co-management agreements position hunters and local communities as responsible agents in resource management, while Greenland primarily pursues resource management on the basis of political negotiations between one national NGO (nongovernmental organisation) and governmental departments. Greenlanders manage their extensive rights to self-determination under Home Rule in a very centralised fashion, making little space for regional input and responsibilities, whereas Native peoples in Canada and Alaska manage their restricted rights in a decentralised fashion, creating more space for different regional input. The battleground for management conflicts and solutions is thus different from region to region, despite the promotion of a common Inuit voice on TEK in the Arctic Council and in environmental cooperation.

Apart from these institutional and political considerations, cultural perceptions of tradition and community homogeneity play a major role in how the debate about knowledge evolves. In Alaska and Canada, traditional cultural values (including TEK) have a prominent position in the image of the good life and in the process of empowerment. In Greenland, the emphasis has seldom been placed on cultural values and local knowledge, but rather on how Greenlanders could control development. In Greenland, colonisation has primarily been seen as alienated from Greenlandic conditions and the Greenlanders. A different perception dominates the North American Arctic, where the outcome of colonisation has been considered to be cultural destruction and loss. This may be one of the reasons why revitalisation of cultural elements from a perceived past is primarily encountered in the North American Arctic. Since the introduction of Home Rule in Greenland, culture, community and knowledge have not really been politicised to any great extent. Although hunters blame biologists for thinking in Danish ways and demand a voice in research, it has not reached the same political magnitude as in the North American Arctic, where culture and TEK have been important tools in gaining and legitimising self-determination.

On the level of knowledge production about resources in Greenland there seems to be a sharp division between biologists and hunters where the latter have little or nothing to say. Biologists are the recognised expert advisers to the Home Rule administration. However, when it comes to the level of political negotiations, hunters (and pri-

marily their organisation KNAPK) play a significant role in the elaboration of hunting regulations, and biologists and Home Rule managers
often see their advice and ideas overruled. The political negotiations
allow a strong input from the hunters' interest organisation, whose
staff work as professional lobbyists and negotiators. Owing to the centralised political structure where decisions are made in the capital of
Nuuk on the basis of political negotiations, local hunters often have little influence on the elaboration of regulations. Hearing and consultation procedures with local users are pursued infrequently and
sporadically both by the Home Rule administration and KNAPK. Thus
the Home Rule bureaucracy constitutes an important role as a buffer
between different interest groups and perspectives and tries to com-
bine short-term socioeconomic interests with long-term national interests in the resource base. Local involvement and responsibility in
resource monitoring and management is not institutionalised, which
adds to the hunters' sense of marginality and frustration.

In both the North American Arctic and in Greenland there are quite
different ways in which community borders are drawn in the speeches
of leaders, elders and in the discussion on ethical rules for research. In
the North American Arctic, local communities are as undivided
wholes, which must be protected from uncontrolled outside interference (for example, researchers) and which maintain this sense of cohe-
siveness by sharing within the community. In Greenland, the frame is
national and divided among interest groups competing for resources
and political influence.

Generally speaking, the promotion of conceptual and institutional
interfaces to bridge the knowledge gap (for example, between biologists and hunters) in the Arctic has primarily been pursued on the
basis of a promotion of TEK and the integration of local users. The
actual solutions possible and preferable are, however, not so easily
promoted because the solution has to fit into the specific social, cultural and political context. The institutional setting forms the basis for
cooperation, and constitutes the area of interface where knowledge can
be exchanged and reality, as well as concepts, can be negotiated. The
Greenlandic setting is extremely focused on the level of negotiation
between a few stakeholders and is not very open to ideas of devolution
of management, as seen in the North American Arctic. However,
Greenlandic hunters and fishermen have been successful in exerting
influence on the management regime through their organisation and
the general goodwill among politicians. In this respect local knowledge
has not been as relevant as part of an overall empowerment strategy as
in the North American Arctic.

Acknowledgements

The research for this chapter was carried out as part of the 'Environmental Knowledge, Cultural Strategies and Development in Greenland and the Circumpolar North' project financed by the Danish Research Councils (TUPOLAR-project). A number of people have helped to sharpen the arguments in this chapter, particularly Susanne Dybbroe and Andreas Roepstorff (both from Department of Ethnography and Social Anthropology, Aarhus University), Jens Dahl (International Work Group for Indigenous Affairs), and Mark Nuttall and David G. Anderson.

Notes

1   Monetary affairs, foreign affairs, police, justice and military affairs are outside the authority of the Home Rule administration.

2   In Danish, the concept 'de grønlandske betingelser' is used.

3   See Oldendow 1936; Jørgensen 1964; Jenness 1967; Viemose 1977; Sørensen 1983; Dahl 1986b; Fleischer 1998 for an elaboration of the political history of West Greenland.

4   In 1937, for example, the provincial council of North Greenland discussed the negative effects the increasing numbers of motorised boats had on the behaviour and presence of the valued beluga whales (Nordgrønlands Landsrådsforhandlinger 1937: 1153–58). Detailed local knowledge was put forward by the council members but an agreement to decrease the traffic could not be found because motorised boats were considered the foundation for further development.

5   The establishment of the Alaska Eskimo Whaling Commission (Huntington 1989; 1992: 110–15) as a regional Inuit organisation managing bowhead whaling, and the direct involvement of Anguvigaq, a Native organisation in northern Quebec composed of hunters, in the creation and implementation of a beluga whaling management plan (Osherenko 1988: 25–32) are but two examples of decentralisation and allocation of responsibility to Native users in North America. For discussions of knowledge and management in North America see, for example, Collings (1997b) and Freeman and Carbyn (1988).

6   Translated by the author.

7   For an elaboration of the concept of knowledge tradition please see Barth (1995).

8   Sometimes, Greenlandic socioeconomic problems are confronted with the simple suggestion of sending Danes home (see for example, Bjerre and Nielsen 1998a, 1998b).

9   A similar worry is expressed by one Alaskan elder in Nooavik (McNabb 1991: 71).

10 The Alaskan newsletter *Sharing Our Pathways* is a good example of an initiative where knowledge for the good life is shared within and between Native communities.

11 It is the same dichotomy one sees between scientific vs. traditional knowledge.

12 The cleavage between past and present is not solely a Greenlandic phenomenon, but can probably be found in other Arctic communities as well, although Dorais (1996: 31) argues that 'in Canada [with the exception of Labrador], the Inuit see no cleavage between their past and present'.

13 The Danish word used in the report is *henlægge*, a verb which is primarily used about things and food, which are put away for storage.

14 Life in small villages is often talked about with a trace of romanticism (see interviews in Nilsson 1984; Langgaard 1986).

15 Jens Dahl (1986a: 320) also notes the lack of organisation and leadership outside the political system.

16 The image does not necessarily correspond to the actual fragmented communities (see for example, Lantis 1972; Warry 1998: 210), or to the fact that local people actually accept a heterogenous distribution of knowledge within a community (Morrow 1990).

4

Codifying Knowledge about Caribou: The History of *Inuit Qaujimajatuqangit* in the Kitikmeot Region of Nunavut, Canada

Natasha Thorpe

In the Kitikmeot region of Nunavut, Canada, the idea that one species could manage another was once unfamiliar to local Inuit. Traditionally, Inuit interrelated to caribou with respect and reciprocity, using the fundamental tenet that all species are equal and interconnected. For Inuit, whose identity, culture and survival were inextricably bound to caribou, the cultural beliefs, traditions and customs known as *pitquhiit* (plural) described their interrelation to caribou and other animals. This chapter considers the question of whether or not new practices used to codify *pitquhiit* truly reflect its nature. My main argument is that what hunters today in Kitikmeot term a 'caribou code' is indeed a different way of talking about caribou, but that it is a positive adaptation to two hundred years of intense change. I will argue that the current attempts to define 'traditional knowledge' and to build a system of 'co-management' for caribou are active and positive ways that elders can counteract the changes brought upon *pitquhiit* since the eighteenth century.

The material for this chapter is from the Kitikmeot region of Nunavut (see Figure 4.1). The research comes from the work of a local initiative called the Tuktu (Caribou) and Nogak (Calves) Project (TNP), which operated in the region from 1996 to 2001 (Thorpe et al. 2003). The aim of the TNP was to document and communicate some of the rich knowledge of caribou held by Inuit. Over the course of the project thirty-seven elders and hunters were interviewed, from the communities of Iqaluktuuttiaq (Cambridge Bay), Qingauk (Bathurst Inlet), Qurluqtuq (Coppermine or Kugluktuk), Umingmaktuuq (Bay Chimo) and the outpost camp of Haniraqhiq (Brown Sound).[1] Over the five-year span of this collaborative research venture, elders shared legends demonstrating their historical and cultural imperative to honour cari-

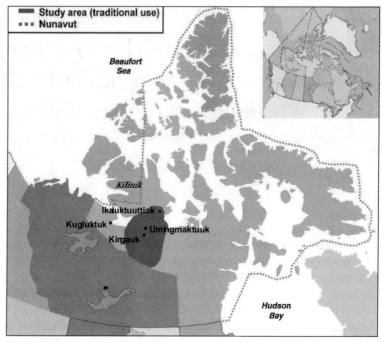

Figure 4.1 The Kitikmeot region of Nunavut, Canada.

bou. These can be understood as a special form of knowledge, illustrating their intimate understanding of the interaction between animals and other ecosystem components.

Should Traditional Ecological Knowledge (TEK) be Codified?

Indigenous peoples have long held an understanding of their landscapes which includes the relationships between ecological as well as social, cultural and spiritual values. This understanding is commonly termed 'traditional ecological knowledge' or TEK in the scholarly literature. The formal definitions of TEK are varied.[2] For the purposes of this chapter, I will assume a broad definition: an accumulated and evolving body of knowledge that comprises the intergenerational survival skills, beliefs, practices, wisdom and experiences of people who demonstrate an acute awareness of dynamic interactions between people, lands and resources. However, it should be stated from the outset that Inuit in the Kitikmeot region understand TEK in a local, hybridised way that I will argue makes it close to the local set of practices known as *pitquhiit*. The most common way of talking about TEK is to use the term *Inuit Qaujimajatuqangit* (IQ) (literally 'the Inuit way of doing things'). According to TNP researcher Margo Kadlun-Jones, *Inuit Qaujimajatuqangit* is the accepted term used to express 'what is known, what has always been known, and what must be known'.

The value, validity and utility of TEK has been a very controversial topic. On the one hand, urban scientists often feel that TEK is nothing more than an unsystematic collection of sayings which has not been formally 'tested' or 'verified'. On the other hand, although Inuit hunters often feel more comfortable with documents that reflect their stories and observations, the codified form in which the knowledge is presented is nonetheless one step removed from the rich way that knowledge is used in everyday life. Despite these contradictions, the movement to document and codify TEK has been growing in the Canadian Arctic. In the last twenty years in Canada, TEK has been given legitimacy through monumental court decisions (e.g., *R*. v. *Sparrow* 1991; *Delgamuukw* v. *British Columbia* 1997), successful land claims and treaty negotiations (e.g., Inuvialuit and Nunavut Land Claims Agreements), the establishment of co-management regimes (e.g., Inuvialuit Joint Commission, Nunavut Wildlife Management Board) and the rise of indigenous influence in formal decision-making structures (Kuhn and Duerden 1996; NSDC 1999; Usher 2000).[3]

Today, community members, researchers and decision makers have moved beyond the point of demonstrating the value of TEK and towards the challenge of finding ways in which TEK can be applied (Usher 2000).[4] In Nunavut, Inuit and non-Inuit organisations are currently discussing how to define specifically Inuit forms of TEK – or IQ – as the first step in any formal process that requires the application of such knowledge (Irlbacher 1997; NSDC 1999; Ferguson pers. comm. 2000; Tigguluraq pers. comm. 2000). Although still striving for an agreed-upon definition four years after the implementation of the Nunavut Land Claims Agreement (NLCA), Nunavummiut, or the people of Nunavut, are striving to define and legislate their unique knowledge, IQ.

Part of the difficulty of defining TEK comes from determining if it is a different way of knowing separate from both urban science and from local ways of acting in the environment. The generic TEK literature often characterises traditional knowledge with a string of adjectives that emphasise its difference from knowledge taught in formal institutions. Adjectives such as aggregating, changing, orally passed, intergenerational, complex (layered), iterative, adaptive and grounded in spiritual underpinnings are some of those used to imply that TEK is different and unique (Berkes 1999; Neis et al. 1999: 220; Wenzel 1999: 114; Thorpe 2000). However these attempts to discriminate a special way of knowing can run the risk of exoticising local knowledge. Critics caution that TEK should not be romanticised as the 'exotic', assumed to be more than it can be nor taken out of context (Cruikshank 1998; Krech 1999). Berkes (1999) and Cruikshank (this volume) speak to several myths that qualify the usefulness of TEK. Key amongst these myths is the idea that indigenous peoples are 'ecologically noble savages' to whom strict norms of behaviour are applied, making them vulnerable to the accusation of being 'Fallen Angels' (Berkes 1999: 145; Cruikshank this volume). Many would like to see these 'exotic' peoples continue to live this way as an example for others; yet when indigenous peoples do not live up to this ideal, they are condemned.

This mythic challenge came to the forefront in the 1990s when Canadian federal and territorial governments took the progressive steps of attempting to form TEK policies (Irlbacher 1997).[5] In response to a TEK policy proposed by the Government of the Northwest Territories (GNWT) in 1993, Howard and Widdowson (1996) argued that it was unconstitutional to have imposed a TEK policy on the grounds that TEK is 'no more than a religion'. Their argument demonstrates a lack of understanding of the spiritual component of TEK and it points to the prejudice against and the suspicion of indigenous knowledge systems (Stevenson 1997; Usher 2000). However, they do suggest that TEK must not be categorically assumed to be valuable simply by virtue of its coming from the 'noble savage'. This is a valid point, given that such blind faith can ultimately be more damaging and disempowering to indigenous people than if TEK were simply rejected as invalid: it can raise unrealistically high expectations of indigenous peoples and their knowledge.

However, such abstract expectations inflate the claims that hunters make on their ways of knowing. Most hunters comment upon their local area and do not tend to generalise beyond the spatial area with which a person is directly familiar. Based on ideas put forth by Cunnison (1951), Cohen (1989: 10) suggests that 'in everyday life, common folk produce and maintain histories of their own little collectivities and resist the construction of more universal historical compositions'. This emphasis on nearby environs may be explained as being part of a cultural norm whereby people are not comfortable speaking about that which they have not seen (Kuhn and Duerden 1996; Krupnik and Vakhtin 1997; Thorpe 2000). During gatherings, Inuit typically do not speak of happenings outside their experience nor are they comfortable speaking on another person's behalf (Briggs 1970; Eyegetok pers. comm. 1999; Kadlun-Jones pers. comm. 1999; Stern pers. comm. 2000). It is more common for hunters to share their personal observations and experiences and compare these with other hunters directly than to 'test' their 'own' knowledge against competing views. These practical aspects of knowing suggest that there can be no clean boundary between what is known and acted upon locally, and other ways of trying to understand the world. Thus it is conceivable that particular hunters might try to 'mix' orally inherited knowledge with hypothetical-deductive knowledge of particular scientists whom they deem to have valuable first-hand experience. This raises the prospect of a 'hybrid' form of knowing that is neither traditional nor urban-scientific but something new developed in response to changed conditions (see Nagy this volume).

One area in which great difficulties occur in Nunavut is with trying to 'manage' the spiritual underpinnings that closely link people and caribou through an ethic and respect for the natural world (Johnson 1992; Inglis 1993; NSDC 1999; Hakongak pers. comm. 2000). Settlements created barriers between Inuit and the natural environment, thereby 'gradually distancing (Inuit) from (their) immediate physical and spiritual attachment to the land, an attachment at the very core of (Inuit) traditional culture' (NSDC 1999: 4). It is through this spiritual attachment that Inuit extend the same or greater respect to caribou as they do people in a way that contributes to a precautionary approach and conservation ethic (Gunn et al. 1988; Johnson 1992; Kofinas 1998; Berkes 1999; Krech 1999; Thorpe 2000; Thorpe et al. 2003).

As TEK is taken from the oral to written domain, understanding of ecological complexities and spiritual dimensions can be lost. Throughout my fieldwork, Inuit explained to me that it was difficult to define IQ or TEK in general because many Inuinnaqtun terms do not translate accurately into English (Atatahak pers. comm. 1999; Maghagak pers. comm. 1999). Writing a definition down 'will never be an adequate format for the teaching of indigenous knowledge' (Berkes 1999: 28), as could be said for all teachings. In other words, a young boy will learn more from watching and listening to his grandfather hunt a caribou and then experiencing the hunt himself, than by reading a book about hunting. This was the case for Alootook Ipellie, an Inuk[6] who speaks to TEK as a unique way of knowing:

> I was often in awe of the extraordinary abilities of my elders to understand the season, in knowing the behaviour of all Arctic animals species and to co-exist with their fellow Inuit in a common goal to survive as a

collective. In the Arctic's harsh environment, one mistake or a lapse in judgement could spell certain disaster. By observing, listening and prac- tising what my elders did, I was instilled with the will to survive for the moment and go on for another day. (Ipellie 1997: 98)

Gerry Atatahak puts forth an Inuinnaqtun perspective on TEK, based on his experience of running the Naonayaotit Traditional Knowledge Study (NTKS), an ongoing project in the Kitikmeot region. He states that 'TEK is *engilgaat* (long ago) *elihimayaghait* (what they should know)' (pers. comm. email 2000). Inuinnaqtun instructor, Margo Kad- lun-Jones, in reviewing this definition and chapter, added that the word *elihimayaghait* is stronger and is more accurately translated as 'what they must or have to know' (pers. comm. 2001). Atatahak explains that 'this information has been passed down through the gen- erations as what [Inuit] know from their fore fathers or mothers and it is a way of still doing things' (pers. comm. email 2000).

Despite the considerable evidence demonstrating that the process of defining or codifying knowledge changes it in a fundamental way, the Government of Nunavut (GN) insists on first defining TEK in an Inuit context before integrating it into everyday policies, procedures and educational programmes. Leading the way in this regard is the *Inuit Qaujimajatuqangit* Task Force (IQTF), run through the Department of Culture, Language, Elders and Youth (CLEY). The IQTF is comprised of six people: two GN employees, two Nunavut Social Development Council (NSDC) members and two elders.[7] In addition to the IQTF, each of the ten departments in the GN is supposed to have an IQ com- mittee, although not all departments have achieved this yet. The chal- lenge for the departmental IQ committees is to provide guidance and to ensure that IQ is incorporated within each department. In the mean- time, the IQTF has put forth an IQ policy that contains, among other things, a definition for Inuit TEK as simply *Inuit Qaujimajatuqangit*. This policy is currently waiting to be approved first by the Nunavut Cabinet and then by the Nunavut Legislative Assembly (Quassa pers. comm. 2001).

The very practical way that Inuit today assess knowledge, and the bureaucratic need to codify knowledge, may help us to move beyond the two dichotomies that are constantly used to criticise TEK. While it may be true that documenting the informal way that elders practise *pitquhiit* is a new or even poorer reflection of the way that knowledge might have been communicated out on the land or by the fireside, it is also true that elders are responding to a different political ecology of self-governing institutions, new forms of communication, and changing hunting technology. Rather than condemning TEK as a pale reflection of real *pitquhiit,* my approach has been to consider how talking about caribou, and if necessary 'codifying' norms related to caribou, helps Inuit hunters to gain an important stake in the quickly evolving world of land claims and co-management negotiations that characterise Nunavut today.

The Caribou Code

Part of the process of defining and legislating IQ has been directly involved with the management of the Bathurst caribou herd through the newly established Bathurst Caribou Management Planning Committee (BCMPC).[8] Still in the early stages, the BCMPC is seeking ways in which to incorporate IQ of caribou. As documented through the TNP, hunters and elders have developed a 'caribou code', which represents some of the most important facts about caribou that need to be reflected in official documents and management plans. In the excerpts that follow, I have selected statements from authoritative individuals which elaborate the 'caribou code' governing behaviour of people that are to be respected. Statements such as these are considered *maligaghat,* or rules, in that they must be adhered to since they are necessary for the good of everybody. In the interpretation following the statements I will argue that these statements represent more than a mechanical conflation of behaviour and norms, or the use of 'spirituality', but instead give highly localised examples of ethically proper behaviour which encourage the listener to think about his or her behaviour.

A hunter must not to be too particular when hunting caribou.
If you ignore the first caribou you see and you do not shoot it then you are not going to have good luck after that. You should not be choosy when you are out caribou hunting. That is one thing that I have heard...The story I told earlier was that you are not supposed to just bypass the first caribou you see hoping to see a bigger or better one or a bigger one. I just bypassed that one caribou and I said we might see more today. Sure enough, we did not see any more caribou all day long. Some things are true. Especially when the caribou are not plentiful, then you try not to ignore the first you see. If you really want it, you will get it. Otherwise, if you just ignore it, you will not see any more. (Hakongak 1998)

Caribou death must be as painless as possible.
Do not let the caribou suffer ...shoot it. You do not let it run off. You chase after it until you shoot it down and never let it go. That is what we do. Even if it is ten miles away, we still go after it. Make sure that it is not left to suffer and die somewhere. Like my brother, when he goes hunting ...he makes sure if he wounds a caribou, (that) he gets it...Except only if he does not have enough gas or bullets to. Then you do not have a choice and you have to make your way home too...If we see a wounded caribou just hobbling or just laying there without moving except to look around, we shoot it because we know it is sick or wounded . . .We would rather shoot it than let it suffer. (Anonymous 1998c)

Caribou meat should not be wasted.
The traditional ways of hunting are not being used any more. Sometimes when people catch caribou, they do not use the whole caribou. It is not acceptable that way. That is not so good. Sometimes without using up all the caribou, part of it would be thrown away. In the olden days, they would use the whole caribou. They never threw anything away when they caught caribou back then. (Maniyogina 1998)

A hunter must share his/her caribou harvest.
We send the meat out to family all the time for the ones who do not really go out and have drymeat to eat...It is not only for us that we shoot the caribou. We always give out drymeat, pack it up and send it out and call to make sure somebody gets their meat before it gets rotten. So it is not only for us that we hunt caribou to make drymeat. It is for family all over...(If you do not share your meat) nobody is going to like you. You are so greedy, get out of my cabin! You are not (going) to have any of mine ...I do not want to know you! (Anonymous 1998d)

A hunter must not boast about his/her harvest.
It is the same with all the other animals, they can hear you talking no matter how many miles away or you are in the house. All animals, they listen to you talking, they hear you talking. I have heard from the old folks that one person was saying how great a hunter he was, very great hunter. He really knows how to hunt, bragging, bragging all the time. Then one day he said he was really sure of himself. He was going to get an animal. He was going to shoot a caribou. He said 'I am gonna go get a caribou from over there.' So he went over there, hunted all over the place, cannot find anything. . .Came back home with nothing. And people kept telling him that he should not brag about these things, otherwise the animals are going to hear you and they know exactly where you want to go. They are going to move away from you and nothing is going to be there. Everything is going to hide from that area. (B. Algona 1999)

Caribou must be respected as spiritual beings.
I can remember one time my mom saying when you catch a pregnant
cow in the spring you are supposed to take the foetus out and put snow
in its mouth as water for the after-life...Put snow in its mouth for its first
taste of water or something fresh other than mother's milk. Supposed to
give you good hunting luck too. Supposed to be able to always have
good luck after that. (Hakongak 1998)

These quotes speak not only to the centrality of both spirituality and
ethical behaviour in following *pitquhiit*, but also to a kind of moral
suasion. Further, such *pitquhiit* are grounded in stories of example and
personal experience that give them more credence and relevance. The
consequences of not following the caribou code can be dire: nobody
will like you, the animals will hide from you or you will have bad
hunting luck. Necessarily, it follows that 'good' behaviour, defined by
adhering to the code, will bring popularity, an abundance of animals
and good hunting luck.

The way that caribou *pitquhiit* worked forty or even ten years ago is
arguably different than the very rationalised form of this written 'caribou code'; however, the interrelationships between living *pitquhiit*
today and IQ are complex. It would be a mistake to understand these
paper restrictions as simple injunctions. Rather they can be better
viewed as strategic positions which are an active response to the rapidly developing administrative climate of Nunavut today. In order to
understand the eloquence of the code, I wish to summarise three primary changes that have challenged the way that *pitquhiit* were reproduced in the recent past. Knowing this context, the attempts to
produce a caribou code, along with attempts to manage the Bathurst
caribou herd, can be seen as proactive attempts to reinvigorate
pitquhiit in twenty-first-century conditions.

Factors Influencing the Reproduction of Caribou *Pitquhiit*

Recognising that this list is not exhaustive and that factors of change
are complex, there are three primary factors responsible for the
changes seen in the depth and distribution of caribou *pitquhiit* in
recent generations. The negative factors include: the influence of new
media technology, changes in hunting technology, and interventionist
state policies. The positive factors include the documentation and production of IQ.

New Media Technology

In the Kitikmeot region of Nunavut, the influence of non-Inuit culture began in the late 1700s with explorers searching for the northern passage and continues today with the presence of many non-Inuit living and working in the North. Further, advances in communication technology have provided unprecedented conduits for urban ways to influence Inuit communities, especially in the last decade. Inuit who once used hand-held radios considered these to be on the leading edge of technology. Now, people are able to connect with one another as well as the rest of the world with satellite phones. This means that community members of Umingmaktuuq and Qingauk, all of whom live without running water and electricity, now have the capability to access the global environment via the Internet.

As the symbol of our information age, television is one of the main ways in which Western culture is transmitted to Inuit. Its importance can be best illustrated through the following story of a modern-day hunting camp. During one of the first years that I lived in Umingmaktuuq for the spring and summer, I was invited to go with a family to a fishing camp about 15 km south of the community. The ocean was still frozen and safe for travel so we loaded up sleds and snowmobiles with fuel, tents, food and fishing line. Together, our group totalled sixteen and spanned four generations.

It was not long before we were settled into our canvas tents at this traditional camp, a stunning ocean-side paradise. The elders told stories about how this had always been used by Inuit because of the good fishing and the flat and sandy campsite. On the second day of our trip, one of the teenage boys was sent back to Umingmaktuuq for something. As he was instructed in Inuinnaqtun, which I did not fully understand, I assumed he might be fetching more supplies. A few hours later he returned with a full sled – complete with a generator, television and VCR. The next thing I knew, people had dropped their fishing rods, started the generator and were huddling inside one of the canvas tents to watch the latest action thriller movie.

As a young researcher new to Nunavut, the juxtaposition of the 'serene and pristine' Arctic environs and the thunderous heaving of the generator was quite a sight and sound to behold. It taught me, however, just one of the ways in which industrial media are influencing Inuit culture and how modern hunting camps are radically different from those described by elders decades ago. Whereas the elders were sharing traditional stories and showing various techniques for preparing fish, most of these lessons disappeared when 'Hollywood' arrived.

Today, the modern hunting camp is a place to 'get away' or to 'get out on the land' rather the way of life. However, this does not mean it is an escape from modern influence and technology. Instead, new technologies may challenge old ideas, but they do not eliminate knowledge.

Many community members feel that the greater the influence of industrial culture, the less room there is for traditional Inuit culture. While overlap in the two cultures exists, there are also areas of mutual exclusion. As it becomes easier for industrial ways to exert a significant influence, it necessarily grows more difficult to preserve *pitquhiit*. As Krupnik and Vakhtin (1997) suggest, perhaps *pitquhiit* are being lost either in purpose, or practice, or both. Instead, older Inuit traditions are replaced by a hybrid paradigm of the old and new ways.

Changes in Hunting Technologies

At the same time that cultural modernisation has redefined hunting camps and reduced the significance of caribou subsistence for Inuit, the introduction of new weapons and transportation technology has made hunting easier and more efficient and thereby altered the ways in which Inuit hunt (Riewe and Gamble 1988; Wenzel 2000). The use of rifles rather than bows and arrows, and snowmobiles instead of dogteams, has changed how Inuit track and harvest caribou. (See Figure 4.2)

Figure 4.2 An impression of the modern landscape of Umingmaktuuq (Bay Chimo) by Logan Kaniak (eight years old) in 1997.

When the first outpost stores opened, hunters quickly bought rifles to replace their bows and arrows. By the 1920s, the bow and arrow had become a relic of the past for the youth (Jenness 1959; Rasmussen 1929 in Diubaldo 1985) and hunting became more efficient with the introduction of rifles. Previously, Inuit built long rows of stone cairns made to look like people, or *inuksuit*, that would steer the caribou towards the waiting hunters (Akana 1998; Alonak 1998; Kuptana 1998). With a bow and arrows in hand, these hunters would crouch behind large rocks that served as hunting blinds called *talut*. The best hunters waited until two caribou were walking parallel, so as to release just one arrow to the necks of them both (Alonak 1998). As soon as the caribou came close enough for the hunter to take aim, the sky would rain arrows. Engineering this cooperative effort was not a straightforward feat. Indeed, the art of coordinating all actors on the hunting stage evolved over generations. As weapons technology improved, the act of hunting became more of a process than an art. The long range of the rifle meant it was easier for hunters to hunt alone and to be more effective in both hitting and killing their target.

The snowmobile was another key factor in advancing hunting technology. Travel by snowmobile is much faster than by dogteam, such that hunters spend less time travelling the land to search for caribou. Especially when hunters travel as a group, they are much more effective because individuals can fan out to look for caribou tracks, and once the tracks are spotted, hunters can quickly follow these towards the caribou. As has been observed in other aboriginal communities today, people more often hunt using numerous short trips of a few days' duration than of the traditional long trips (Berkes et al. 1995; Hakongak pers. comm. 2001). As a result, hunters spend less time on the land and more time in communities (Berkes et al. 1995; Condon 1995; 1996; Riewe and Gamble 1988).

In the past, hunters consistantly needed to feed their dogs to keep their teams 'fuelled'. When the caribou herds shifted away from Inuit camps, it became more difficult to feed the dogs (Jenness 1959; Akana 1998; M. Algona 1999; Kamoayok 1998; Kaniak 1998). As the dogs became malnourished, the strength and stamina of the team waned and so it became impossible to travel great distances in search of caribou. Such hardships sometimes meant that Inuit starved to death.

Since snowmobiles can race over the land at speeds far greater than those achieved by dogteams, hunters can travel great distances and expand their hunting grounds. Limited only by the size of their gas tank or capacity to carry extra fuel, hunters can travel as far as necessary to find caribou, even when migration routes of the herd shift away from communities (Riewe and Gamble 1988; Hakongak 1998; B.Algona 1999). This reduces the variability of caribou prevalence. The subsequent predictability of supply necessarily decreases both the demand for and value of caribou.

The hardships brought by scarcity of food and poor conditions for hunting are much less likely today due to the introduction of store-bought foods, changes in hunting technology and the establishment of settlements. Accordingly, the demand for caribou has fallen while weapons and snowmobiles have radically changed hunting ways and continue to contribute to the fading of *pitquhiit*. With the snowmobile able to bring the hunter faster and farther on the land and the rifle extending the hunting range, it is easier and usually more fruitful to hunt caribou. Diminished are the requirements for such *pitquhiit* as patience and stealth, once critical when hunting was a cooperative effort. Thus not being particular about which caribou you hunt or standing down-wind of the caribou are less relevant than formerly, since the hunter is at a greater advantage today than in traditional times.

State Management

The advent of state wildlife management policies and practices grounded in Western values has largely replaced Inuit practices based in *pitquhiit*. Whereas the pervading Western system emphasises capitalism, individualism, hierarchy, competition and centralised authority, the original Inuit system was traditionally influenced by cooperation, sharing and working together for the good of the group (Freeman 1985; Riewe and Gamble 1988; Shapcott 1989; Hunn 1999).

The implementation of state policies, particularly in the 1940s, effectively assimilated, 'civilised', acculturated and exerted control over Inuit. This led to the gradual attrition of traditional *pitquhiit* (Tester and Kulchyski 1994; Berry 1999; NSDC 1999). With concerns about sovereignty, the previous peace-time laissez-faire policies were replaced by postwar interventionism (Usher, this volume). Inuit children were plucked from their traditional outdoor 'classrooms' and placed in centralised educational institutions, while entire families were moved from their traditional territories to defensively strategic locations, often poor in hunting potential.

The 1950s were marked by wildlife biologists who criticised Inuit hunting practices as 'savage' and 'wasteful' (Kelsall 1968; Riewe and Gamble 1988; Cizek 1990; Kofinas 1998; Kendrick 2000). As Inuit lost access to their land, language and traditional way of living, *pitquhiit* that formerly governed wildlife 'management' were necessarily set aside by imposed state policies aimed at limiting waste of caribou meat (e.g., feeding it to dogs), curbing aboriginal harvests, increasing education and enforcement of hunting regulations and restricting the use of caribou meat at institutions (e.g., schools and hospitals). Thus, the federal stance of maintaining sovereignty was manifest in the relocation, immobilisation and supervision of Inuit (Duibaldo 1985), which left little room for local input into state management of caribou (Kofinas 1998).

As I have been told frequently, Inuit were scared of offending or disobeying state officials, particularly those in enforcement roles. Still, small acts of rebellion were staged including the killing of caribou and muskox when the practice was discouraged, especially during times of need (Diubaldo 1985; B. and P. Omilgoetok pers. comm. 1998; Berry 1999). Generally, however, whereas Inuit used to make decisions important to their relationship with caribou, they now relinquish their ownership of this process to the state. This occurred in the form of hunting regulations, quotas, licences and fees for violations (Osherenko 1988; Riewe and Gamble 1988).

From the 1960s to the 1980s, many scientists working in Nunavut came to appreciate that Inuit have had important contributions to make to an understanding of caribou (Gunn et al. 1988). Successful examples had been demonstrated in Greenland and Alaska where 'the responsibility for the management of wildlife (had) been turned back to the people' (Bourque 1981; Clausen 1981: 4; Davis 1981).[9] Such models were now beginning to influence state policies and practices despite widespread distrust and scepticism. Meanwhile, negotiation of the NLCA began, providing momentum and empowerment to Inuit in making their views instrumental in wildlife management.

Recognising that data from both biologists and Inuit were needed to make sound decisions and that efforts were required to bring these seemingly disparate views together, several joint management bodies emerged. In 1982, the Beverly and Kaminuriak Caribou Board was formed in response to the 'crisis' of perceived declines in the herds (Freeman 1985; Cizek 1990; Kendrick 2000). In 1985, the Porcupine Caribou Management Board formed partly due to conflict over oil and gas development in the Beaufort Sea, and on the range of the caribou herd in Alaska and the Yukon Territory (Kofinas 1998). Both of these boards have membership by government and indigenous peoples. Ironically, the development of such organisations is in keeping with tenets more typical of traditional Inuit ways, namely, cooperation, sharing and working together. In bringing indigenous and nonindigenous groups together, the boards were able to steer away from state governance models based on counterproductive goals such as hierarchy, individualism, competition and capitalism.

Signed in 1993 and implemented in 1999, the NLCA was monumental in its provisions to incorporate IQ wherever possible,[10] to preferentially hire Inuit in the new GN and to establish three co-management bodies: the Nunavut Impact Review Board, the Nunavut Water Board and the Nunavut Wildlife Management Board. Through these co-management boards and the GN, Inuit could now meaningfully regain their decision-making power and embark on de facto self-government[11] that more faithfully reflected Inuit values and perspectives than previous governance (Légaré 1998; Hicks and White 2000).

One concrete example of the successful adaptation of traditional *pitquhiit* to the new political context is the Nunavut Wildlife Management Board.[12] Prior to the NLCA, wildlife management had been primarily centralised and run from the federal government based in Ottawa, with few responsibilities delegated to the territorial government in Yellowknife (GNWT). The creation of the Government of Nunavut and the Nunavut Wildlife Management Board (NWMB) allowed for decentralisation and a more environmentally, socially and culturally relevant approach to wildlife management. Despite the fact that the ultimate responsibility remains with the federal government, the NWMB is now the main instrument of wildlife management and the main regulator of access to wildlife. One of the objectives of Article 5 of the NLCA, outlined in Section 5.1.3, is:

> the creation of a wildlife management system that is governed by, and implements, principles of conservation; fully acknowledges and reflects the primary role of Inuit in wildlife harvesting; serves and promotes the long-term economic, social and cultural interests of Inuit harvesters; as far as practical, integrates the management of all species of wildlife; invites public participation and promotes public confidence, particularly amongst Inuit; and enables and empowers the NWMB to make wildlife management decisions pertaining thereto.

These objectives are in keeping with *pitquhiit*, particularly with the commitment to conservation and integrated management of other wildlife. If wildlife management is to adhere to these objectives and remain meaningfully powered by Inuit interests and needs, then it remains likely that management will include *pitquhiit*. Whereas the relationship between Inuit and caribou used to be based on *pitquhiit* alone, it is now grounded in a hybrid of state management and traditional Inuit philosophies. In the last few years, a key organisation has developed through the NWMB that may turn out to be a leading example of how *pitquhiit* and state policies and practices can be integrated.

The NLCA established the framework for merging *pitquhiit* and state wildlife management policies. There are additional opportunities for Inuit to regain some elements of control in the management of caribou and other wildlife through the GN, NWMB and local initiatives such as the BCMPC. It is encouraging to note that of the three main factors contributing to the erosion of caribou *pitquhiit*, the direction of state management policies is the single factor that can be controlled most easily and directly. However, while the NLCA sets a template for cooperation, it is yet to be seen whether stakeholders can meet this challenge in a way that is satisfactory to Inuit.

The Formal Documentation of *Pitquhiit*

While recognising that there are many forces working together against the continuance of caribou *pitquhiit*, there are other factors that help to preserve them. As previously discussed, the fact that governance in Nunavut must now incorporate IQ wherever possible necessarily means that efforts to document IQ have increased. Accordingly, *pitquhiit* are at least recorded if not practised.

Recording and documenting IQ across Nunavut is occurring in various ways – through formal processes, projects or societies or as an incidental side benefit – by a growing perspective in government, education and communities in general. Although available funding and personnel cannot possibly meet all of the hopes and expectations of elders and other community members, significant progress is being made (NSDC 1999).

In the Kitikmeot region, a prominent contributor to the gathering of IQ, and the consequent preservation of *pitquhiit*, is the work of the Kitikmeot Heritage Society (KHS). Many local oral history and IQ projects in Ikaluktuuttiaq are supported, carried out or orchestrated by the KHS. Other formalised IQ efforts are ongoing in nearby Kugluktuk, Kuugaruk and Gjoa Haven. Some of these are the Tuktu and Nogak Project (TNP) and the Naonayaotit Traditional Knowledge Study (NTKS), which were both supported by the GN among other agencies. This governmental support is testament to the current trend towards Inuit cultural revitalisation and the integration of IQ into more and more government activity. Departmental meetings, agency mandates and the new co-management boards are all increasingly incorporating IQ – or at least establishing processes for doing so (NSDC 1999).

Similarly, awareness and integration of IQ is growing at the community level. An increase in various elder-youth events (e.g., camps, sewing classes, hunting expeditions) and the development of IQ-inclusive school curricula are some good examples of this growth. Such initiatives continue to be strongly supported by designated Inuit organisations (e.g., Kitikmeot Inuit Association), hunters' and trappers' associations and other community agencies.

Figure 4.3 George Kapolak-Haniliak and his son, Kevin, Bathurst Inlet, Canada, spring 1998.

Activities that bring elders and youth together are essential to the preservation of *pitquhiit* because they encourage sharing of both the purpose and procedures behind *pitquhiit*. In particular, elder-youth camps have proven to be a critical way in which traditions are shared, especially for caribou (Thorpe 1998; Thorpe and Eyegetok 2000a; 2000b; GeoNorth et al. 2002). Such camps have been held by the Gwich'in (Kritsch 1996), Inuvialuit and Inuit (Hakongak 2000; Maltin 2000; Willett pers. comm. 2000; Bromley pers. comm. 2001) and are becoming popular ways to promote cultural heritage (see Figure 4.3).

Together, these IQ efforts are invaluable to the continuance of caribou *pitquhiit*. However, while these events suggest an encouraging and increasing trend towards preserving Inuit heritage, currently there is a relative paucity of IQ that has been recorded or integrated when compared with the remaining need for its documentation. Many of the necessary steps to preserve and revitalise IQ and *pitquhiit* have yet to be achieved (NSDC 1999). There will continue to be an ever-increasing urgency to do more before many elders (or so-called 'superexperts') pass away, carrying with them much of their knowledge.

The factors for and against the retention of caribou *pitquhiit* are not in perfect balance. The infusion of modern markets, trade and technology into Inuit culture, the loss of elders who were raised on the land, advances in caribou hunting technology and the imposition of state wildlife management practice and policies currently outweigh IQ reculturalisation efforts. The balance in favour of factors driving change is such that the gradual loss of caribou *pitquhiit* continues through two primary means: the loss of Inuinnaqtun and the diminished reliance on caribou by Inuit. These are the two dominant outcomes in the gradual redefinition of the relationship between Inuit and caribou.

Pitquhiit *and Caribou Management of the Future*

The alternatives available in modern times mean that caribou are no longer the defining element of Inuit subsistence. Although some elders depend upon traditional foods for up to 90 percent of their needs, the younger generations eat a majority of store-bought goods. Inuit are no longer reliant on caribou and therefore it is not as critical to know or practise *pitquhiit* today as in traditional times.

In the past, a hunter using a dogteam could not afford to be choosy in selecting a caribou nor could he waste meat for fear that his family might starve. Now, with easier hunting through better weapons and faster transportation and the accessibility of larger hunting grounds, Inuit can be fussy, wasteful and fail to share their harvest without perishing:

> Now (Inuit) leave the (caribou) foetus with the rest of the remains when they bring a caribou home. Now, to eat it, it is probably politically incorrect to do that! All that has to do with politics. It is just a difference of how people treat animals now…They can go to the store and buy food. In the old days, when they caught caribou, they used everything, everything on it. (Hakongak 1998)

Of course, the fact that Inuit do not have to rely on caribou for subsistence has implications for Inuit sustenance (Davis 1981; Condon et al. 1995). Based on traditional times, there is still prestige and respect associated with hunting that is misplaced when these hunters enter into the wage economy.

The relationship between Inuit and caribou is radically different today than in traditional times. Changes in hunting technology and communications media and interventionist state policies have together driven changes to *pitquhiit*, despite concomitant increases in efforts to codify a body of knowledge known as IQ. Such changes to *pitquhiit* have altered the relationship held by Inuit for caribou. The ultimate consequences of such changes include the loss of Inuinnaqtun and a lack of reliance on caribou in Kitikmeot communities. Without language to communicate traditional beliefs, thoughts and practices, youths cannot learn from their elders nor foster an interest in *pitquhiit*. These realities are exacerbated by the fact that Inuit no longer rely on caribou for culture, identity and subsistence. Instead, store-bought food is readily available and hunting wildlife is faster and more productive. Consequently, Inuit share or practice *pitquhiit* less and less such that IQ of caribou erodes.

When considering the factors driving change discussed in this chapter, it becomes apparent that we have the greatest power to alter the effects of state management policies and practices. The forces of cultural modernisation are both strong and inevitable, elders will continue to pass away and advances in hunting and communications technology will go on. However, state management is the factor of change that is most easily modifiable in local ways and can be incorporated as long as there is the will and continuing support for the revitalisation of Inuit culture. The NLCA, in combination with a strong commitment by stakeholders to preserve, enhance and communicate Inuit culture, sets the necessary conditions for wildlife management in Nunavut to return towards its roots in *pitquhiit*.

If *pitquhiit* are given more weight in caribou matters, then fewer interventionist state management policies, practices and philosophies could be applied, but this challenge has yet to be fully realised. There is overlap between *pitquhiit* and state management such that they are not mutually exclusive, for example, in the areas of conservation. However, typical state management policies do not emphasise that caribou should be respected or that there is a spiritual dimension between caribou and Inuit. Indeed, there is no room within current systems to accommodate this 'different' knowledge. The challenge for governance in Nunavut is to redefine wildlife management in a way that might allow a more common existence for *pitquhiit* and state policies.

Achieving a balance between the traditional *pitquhiit* and state management involves three steps. The first is to recognise that the forces against *pitquhiit* in the current caribou management regime are significant. Second, stakeholders must move beyond discussion and follow through with their commitments to cultural and heritage efforts to document IQ, for example by providing funding and training. The third stage requires that relevant *pitquhiit* be incorporated into the caribou co-management regimes through a system of mutual respect and open-minded thinking.

In the Kitikmeot region, management of the Bathurst caribou herd currently faces this third stage. The Bathurst Caribou Management Committee has an unprecedented opportunity to incorporate *pitquhiit*. However, with this opportunity comes the responsibility of pioneering a co-management regime that satisfies both Inuit and non-Inuit stakeholders, by meaningfully incorporating and applying *pitquhiit*. This daunting task continues to challenge most agencies charged with finding a common dialogue between Inuit and western cultures – the common ground between IQ and conventional science. As Inuinnaqtun is lost to English and caribou become less important owing to the trends of Western culture and technology, northerners may have to find a new language so that crucial parts of the caribou code are preserved. There is room for a hybrid of English and Inuinnaqtun, a marriage of state wildlife management and traditional caribou *pitquhiit* – but we must have *quinuituq*, deep patience.

Acknowledgements

I am grateful to the elders and hunters of the Kitikmeot region for their patience, courage, grace and generosity of spirit in sharing with me in so many ways. *Takupkaqtarma tariuryuamik ainnikkut hikumi.* This work would not have been possible without my coaches and critics, Sandra Eyegetok, Naikak Hakongak and Margo Kadlun-Jones. Deep gratitude is also given to David Anderson, Mark Nuttall, Dyanna Riedlinger and Shari Fox, who each provided such thoughtful insight. Many thanks are extended to members of funding agencies that made possible the Tuktu and Nogak Project and other IQ works. In particular, the West Kitikmeot/Slave Study Society, Government of Canada, Government of Nunavut, Nunavut Wildlife Management Board and BHP Billiton Ltd.

Notes

1   There are numerous spellings for places throughout Nunavut and reaching agreement on their spelling is near impossible. This is partly because every visitor (e.g. missionary, explorer, Royal Canadian Mounted Policeman) claimed a different spelling and this spelling was passed on to resident Inuit. To help alleviate spelling variations, the Government of Nunavut has supported a standardised system which I follow throughout this paper. For more information about the language debate, see Harper (2000).

2   While there is a rich nomenclature debate surrounding the term 'traditional ecological knowledge', this is beyond the scope of this paper. For ease of consistency, 'traditional ecological knowledge' is used to refer to what has also been termed aboriginal knowledge, indigenous knowledge, indigenous ecological knowledge, traditional knowledge, local knowledge, oral history. At the same time, the utility and validity of these other terms is recognised. See Cruikshank (this volume) for a discussion of how TEK is defined, used and abused.

3   (1). *R.* v. *Sparrow*, (1991) B.C.W.L.D. 689 (B.C.C.A), (1991) B.C.J. No. 304 (B.C.C.A). Note:B.C.W.L.D. = British Columbia Weekly Law Digest. (2). *Delgamuukw* v. *British Columbia*, (1997) 3 S.C.R. 1010, (1997) S.C.J. No.108. * (S.C.J. is a QuickLaw database "Supreme Court of Canada Judgments"). (3). Agreement between the Inuit of the Nunavut Settlement Area and Her Majesty the Queen in Right of Canada (Ottawa: Tungavik Federation of Nunavut and Indians Affairs and Northern Development, 1993). (4). The Western Arctic Claim: The Inuvialuit Final Agreement (Ottawa: Indian Affairs and Northern Development, 1984).

4   In Canada, the value and utility of TEK has been clearly demonstrated through numerous works (Spink 1969; Nakashima 1986; Freeman 1992b; Johnson 1992; Kuhn and Duerden 1996; McDonald et al. 1997; Ferguson et al. 1998; Berkes 1999; Krupnik and Bogoslovskaya 1999; Neis et al. 1999; Wenzel 1999; Cruikshank this volume; 1993). There have been efforts to outline the importance of TEK in environmental assessment and management (Feit 1988; Nakashima 1990; Stevenson 1996; Usher 2000). Other TEK work is ongoing throughout the world, in countries such as Greenland (Sejersen this volume), New Zealand (Lyver et al. 1999), Alaska (Krupnik and Bogoslovskaya 1999) and Siberia (Krupnik and Vakhtin 1997; Krupnik and Bogoslovskaya 1999). Others have commented on the appropriateness or the potential of integrating TEK with other knowledge systems (Huntington 1996; Alvaraz and Diemer 1998; Raine 1998; Krupnik and Bogoslovskaya 1999; Nadasdy 1999; Cruikshank this volume;).

5   For example, in 1996, the Prince of Wales Heritage Centre in Yellowknife started a TK Round Table comprised of researchers and members of both communities and governments. This followed on the heels of a call for inclusion of TEK by the Canadian Polar Commission and the International Institute for Sustainable Development. The Brundtland Report, *Our Common Future*, also pointed to the necessity of TEK.

6   Inuk is the singular form of Inuit. Inuinnaqtun is the language and local ethnonymn for Inuit who live in the Bathurst region. Inuvialuit is the local name for Inuit living in the Western Arctic.

7   In March, 1998 the Nunavut Social Development Council held a workshop in Igloolik which was attended by 120 participants, who worked towards the overall goal to 'set the stage for the development of policies and programmes aimed at ensuring that Inuit culture, language, and values are democratically reflected and used in the day-to-day operations of the new Nunavut Government' (NSDC 1999: 2). A report on this

workshop sets out recommendations in the areas of: governing Nunavut; gathering and using IQ; Inuit society; spirituality, shamanism and customary law; land skills, economy and the environment; health and education; and language.

8 The objectives of the BCMPC are:

(a) to prepare a 10 year management plan for the Bathurst caribou herd in the interest of traditional users and their descendants, who are or may be residents on the range of the caribou, while recognizing and addressing the interests of other users and all Canadians in the management of this resource;

(b) to recommend a process of shared responsibility for the implementation of the management plan prepared in (a);

(c) to establish communications amongst traditional users and amongst the parties hereto in order to ensure coordinated caribou conservation and caribou habitat protection for the Bathurst herd' (BCMPC 2000).

9 Twenty years after the call to integrate Inuit knowledge into wildlife management in Greenland was made, whether the state actually and successfully realised this goal is questionable on both institutional and conceptual levels (Sejersen this volume).

10 The NLCA even went so far as to make provisions for the Nunavut Wildlife Management Board to conduct an 'Inuit Bowhead Knowledge Study' and to set aside $500,000 for this purpose (Article 4: Part 5.5.2).

11 This is de facto self-government in the sense that the majority of members of the Government of Nunavut will be Inuit and therefore exert influence similar to that of self-government.

12 Other successful examples of *pitquhiit* being incorporated into the new political regime include the new Government of Nunavut policy that allows for annual 'storm' days as well as sick days for employees who get stranded on the land while hunting or otherwise pursuing traditional activities. With this new policy, employees do not feel pressured to risk their lives in poor weather in order to get back to work. Other successful examples are documented in Thorpe (2000).

5

A Story about a Muskox: Some Implications of Tetlit Gwich'in Human–Animal Relationships

Robert P. Wishart

Introduction

There is a growing body of literature in anthropology concerned with state–aboriginal relations as they centre on issues of wildlife management.[1] These analyses coincide and refer to another, larger literature regarding aboriginal human–animal and human–land relationships.[2] Rather than revisiting these arguments directly, this chapter makes use of the general messages from these bodies of literature while focusing on what I was told by one wildlife officer and some Gwich'in elders about the ramifications of one particular hunting incident, when a muskox was killed by a Tetlit Gwich'in elder from the community of Fort McPherson, Northwest Territories (NWT), Canada.

'Gwich'in' when roughly translated into English means 'one who dwells' (Osgood 1970: 13) but it also refers to a self-recognised unity as a Nation amongst those who live in the area close to the tree-line in fifteen towns and villages in the Northwest Territories and the Yukon in Canada and in eastern Alaska in the United States of America. Gwich'in are the most northern non-Inuit First Nation in North America. Gwich'in is also an Athapaskan language, which has several dialects. 'Tetlit Gwich'in' refers to those who dwell on the Peel River and it also refers to the dialect of Gwich'in spoken there. In academic literature, Gwich'in have also commonly been known by the ethnonyms 'Kutchin' and 'Loucheux.' Gwich'in is now the favoured ethnonym due to the fact that 'Kutchin' reflects a mispronunciation and 'Loucheux' is a French word referring to the shape of their eyes and is now considered by many to be derogatory.

I present my argument in a first-person narrative manner in order to elicit a local way of knowing that my teachers in the field have taught to me.[3] When I first presented a version of this paper at the Twelfth Inuit Studies Conference at the University of Aberdeen I received several questions regarding wildlife biology as it pertains to the muskox.

This chapter is not an attempt at answering such questions, as I feel it would detract from the local messages, but instead it is an analysis of the content and process of sometimes conflicting views of 'the land' as they are represented through speech. It is also an observation about the cross-cultural implications of experimentation with wildlife. However, in response to a few questions I received regarding the history of the translocation of muskoxen into the area where Tetlit Gwich'in hunt and trap, I have included a brief summary of this fascinating story of wildlife management.

The Situation

In late April of 2000 four Tetlit Gwich'in men from the community of Fort McPherson, NWT, drove south along the Dempster Highway[4] into the Richardson Mountains in search of caribou. This spring hunt for caribou was late, and was brought on by the fact that the Porcupine caribou herd had failed to migrate through Tetlit Gwich'in country that spring. Freezers within the town were now bare of meat from the autumn and winter hunts and there was a certain air of desperation about this particular hunt. The men were prepared to travel quite far (about 600 km) in search of caribou. About 30 km after the men passed the point where the highway enters the Yukon, they saw an animal walking across the tundra and they hoped it was a caribou or at least a moose that was 'not too poor' after having made it through the winter. What they discovered was that it was a lone, young, male muskox. The elder amongst them promptly shot it and the others butchered it in rapid order. They continued to travel south and they did eventually encounter three caribou about 400 km from Fort McPherson which they also killed and butchered. Upon their return trip home they ran into an officer (who is a member of the Vuntut Gwich'in from Old Crow, Yukon) of the Yukon Department of Renewable Resources, who was stopping and checking all the vehicles on the road because she had evidence that someone had killed a muskox.

The elder immediately admitted that he had shot the muskox and explained that he was bringing it home to help feed his family and his community. The elder believed that he was well within his rights to do so according to the harvesting regulations which divide the legal land-scape of the Gwich'in Nation.

The Gwich'in Nation is divided into populations who live in Alaska, the Yukon and the Northwest Territories and each has different rights to territory and subsistence harvesting. In Alaska the Gwich'in decided not to sign the Alaska Native Claims Settlement Act of 1971, which extinguished any aboriginal rights (including hunting and fishing) and replaced these claims with a cash settlement and a set of corporate titles which were to be reviewed in 1991. However, Congress in 1987 amended these corporate claims and extended their tenure indefinitely (Morehouse 1988: 6). The Gwich'in in Alaska therefore still maintain their original claim to territory and their aboriginal rights to hunt and fish. However, according to the Alaska government, the Gwich'in are held to the conventions of the Alaska National Interest Lands Conservation Act of 1980, which somewhat remedied the lack of subsistence clauses in the Alaska Native Claims Settlement Act by giving Native people a 'rural resident' subsistence right and a voice in the construction of any new hunting and fishing regulations (Morehouse 1988: 6). At the present time Gwich'in living outside of Alaska (who may have traditional hunting territories within that state) have no aboriginal rights to hunt within the state (Childers and Kancewick nd:14).

In the Yukon, the Vuntut Gwich'in have a final agreement with Canada, which was signed in 1995 and sets aside lands for their traditional uses and for protection by the Vuntut Gwich'in. Any other member from a First Nation must get written permission to harvest renewable resources from these lands. However, overlaps exist with the Tetlit Gwich'in to the east who have always maintained traditional trapping and hunting areas within parts of the Yukon.[5]

In the Northwest Territories a final agreement also exists and the Gwich'in residing in the four communities of Fort McPherson, Tsiigehtshik, Aklavik, and Inuvik have the right to harvest for subsistence anywhere within the Gwich'in Settlement Area. In addition the Tetlit Gwich'in have the right to harvest for subsistence within either the primary or secondary use areas of the Yukon,[6] according to the Gwich'in Comprehensive Land Claim Agreement, signed 22 April, 1992 (Gwich'in Tribal Council et al. 1992, Appendix C: 48). All of these rights do not supersede regulations as they pertain to protected species.

The elder in question was well within the secondary use area, and while he and his contemporaries think and speak of that land differently from the legal description, as being part of Tetlit Gwich'in 'country,' he was correct from the point of view that he had the right to hunt within that area and he was unaware that muskoxen were in any way protected or should be, for reasons which will become clear later. He and his hunting party tried to argue this point, but the wildlife officer disagreed and did her job, confiscating the meat and taking down the personal information about the elder so that her department could later charge him – should they decide to do so – with the unlawful harvesting of an animal classified as specially protected wildlife.

Two months later the officer's supervisor drove into the elder's wife's fishcamp, located at the place where the Dempster Highway crosses the Peel River (a place locally referred to as Eight Miles), looking for the elder in order to return the confiscated meat. His department apparently decided that there was more to gain through a friendly message about the importance of preserving the muskox than there was in a long, drawn-out legal conflict. According to regulations governing cases where charges are not pressed, he was obliged to return the evidence. He also came prepared to speak to anyone that would listen about the regulations concerning muskoxen.

Needless to say, I believe they made the right decision concerning the method of getting the message across, but what I think is really interesting is the actual message and the implications of it. The preceding story brings forth many issues regarding contemporary life for First Nations people in Canada. But what I would like to focus on here is how there are two distinct views of the land and the animals being played upon by the preceding story: that of the Gwich'in elder and that of the Yukon Department of Renewable Resources. I believe I have been lucky to be privy to both of these views, the former by living with this elder and hearing him talk about the situation as he goes about his life on the land, and the latter because it was I that agreed to take this officer to see the elder in order to talk to him and return his meat.

During the brief time I spent with this officer, he told me about how people in the Yukon Department of Renewable Resources were extremely upset about the shooting of one of the "specially protected" muskoxen that live on the North Slope. He explained that these animals are the descendants of a small group of muskoxen that were brought to the area by biologists as an experiment in attempting to reintroduce a vanished species. He further explained that sometime in the past muskox had lived on the North Slope and probably the tundra-covered areas of the mountains, and it was decided by biologists and renewable resource officials of the time to take animals from areas where they are considered to be plentiful and set up a population of protected animals. The reason for doing this, and the continued protection of this group of transplanted animals, is politically complex but he mentioned a couple of times how part of its justification is based on tourism. It was explained to me that tourists driving along the highways expected to see an abundance of Arctic animals and many were complaining that after driving all the way to the Arctic and not seeing a thing they were better off going to one of the national parks in Canada where many large animals can be seen right on the highway.

The fact that these tourists come to the area in the wrong season to see such things as caribou and moose or that they look in the wrong places at the wrong time of day is of little importance. What is important is that they be given the opportunity to see as many large, typically Arctic animals as possible and muskoxen are quintessential Arctic animals that do not migrate as regularly as caribou and do not hide in the brush that grows around the lakes like the moose. They like open tundra where they can be easily observed and they tend not to flee, which allows for good opportunities for amateur wildlife photography. What it comes down to, the wildlife officer reported to me, is that the country is already spectacular but the tourists need to be able to do some wildlife viewing in a situation that fulfils their expectations of pristine wilderness.

What this officer reported to me is a simplified version of his department's stance concerning the history and place of muskoxen in particular and wildlife in general. A brief history of the Yukon muskoxen is necessary at this point in order to aid in the analysis of the Yukon Department of Renewable Resources. There is physical as well as historical evidence that muskoxen were indigenous to the entire coastal plain of Alaska and the Yukon.

Approximately 150 years ago these animals completely disappeared as a result of overharvesting by whalers and commercial hunters (yourYukon, column #89: 1). Any muskox that now lives at least part of its life in the Yukon is the descendant of a population of thirty-four muskoxen taken from Greenland in 1930 and relocated to Fairbanks, Alaska. The original reasons for this translocation were: '(i) to aid in conserving of a species threatened with extinction, (ii) for contemplated experiments in re-establishing the muskox as a native animal in Alaska, and (iii) for experimentation with a view to their domestication' (Smith 1989a: A23).

During subsequent years, these muskoxen were released on Nunivak Island, Alaska, where they increased their population to 750 animals by 1968 (Smith 1994: 2). In 1969 and 1970 a total of sixty-four (Reynolds 1989: A26) of these Nunivak Island muskoxen were then transplanted by the Alaska Department of Fish and Game with assistance from the United States Fish and Wildlife Service to the Alaska mainland near Barter Island on the north coastal plain (Pederson et al. 1991: 7). These original sixty-four animals have since increased their numbers to an estimated 1998 population of 700. In 1984 the first female muskox was observed in the Yukon and an approximate 150 of

these North Slope Alaskan muskoxen have now established themselves within Canada. Currently there is a population growth of 10 to
15 percent per year although there has been a recent drop in the population over the last two years (Wildlife Management Advisory Council
2001: 1); similar declines were observed in Alaska after an initial population growth and prior to a continual growth in the overall population (Reynolds 1989: A28).

The Alaska Department of Fish and Game considers 'the return of
muskoxen to Alaska [to be] an important success story in wildlife conservation' (Smith 1994: 2), as the original population of thirty-four animals from Greenland have thrived and through translocation and
natural dispersion have claimed many new areas and grown to a population of 2,200 'free-ranging animals' (ibid.). It is hoped by this
department that the population continues to grow and that they continue to spread to new areas because of the potential value to Alaskans
from sport-hunting fees, meat, wool, and because 'this hardy survivor
of the ice ages is an important attraction to tourists, photographers,
researchers, and students of wildlife' (ibid: 2–3).

In the Yukon, muskox are considered to be a "Specially Protected
Wildlife Species" by the Yukon Wildlife Act (1981), one of three mechanisms[7] for determining species 'at risk'. The Yukon Wildlife Act, as
one of its mandates, serves to identify species, that are at risk in the
Yukon, but not elsewhere; the muskox fits into this category. (Yukon
Department of Renewable Resources 2001a: 1). Why muskoxen are
now moving into the mountainous regions (such as that where the
Tetlit Gwich'in hunt caribou) is currently being researched and a 'scientific' explanation is apparently forthcoming. However it should be
noted that Peter Lent (1999: 71), in his exhaustive search of available
materials on the archaeology and the ethnohistory of muskox use in
the circumpolar north, argues that the only Athapaskans to regularly
hunt muskoxen in the past were the Chipewyans. Some evidence
exists that 'Chandalar Kutchin' would occasionally run into muskox in
the northern foothills of the Brooks Range, close to the Arctic plains,
but no mention is ever made of the more eastern Gwich'in (like the
Tetlit Gwich'in) ever having hunted them. My research into the oral
history of Tetlit Gwich'in has also found no evidence for the presence
of muskoxen in these areas.

The Yukon muskoxen are thus a group of animals with an international history of resource management and the Yukon Department of
Renewable Resources is taking its role in ensuring a continued presence of these animals on the North Slope very seriously. The fact that
the presence of animals such as muskoxen may have multiple financial

benefits to the government is not lost on these regulating bodies. Just as in the quote from the Alaska Department of Fish and Game cited above, the Yukon Department of Renewable Resources pays attention to the considerable potential benefits to tourism that an icon of Arctic wilderness, like the muskox, provides.

Muskoxen viewing holds 'special appeal to Firth River rafting parties and visitors to Ivvavik National Park and Herschel Island Territorial Park' (Wildlife Management Advisory Council 2001: 2). Furthermore, the Yukon Department of Renewable Resources dedicates considerable resources to its 'wildlife viewing program'. On its Internet site, the Department of Renewable Resources has 'wildlife viewing' as one of its main search categories – presumably in response to the fact that the most frequently asked question is about potential wildlife viewing. Each month this department also puts out a calendar of special events highlighting the many 'wildlife viewing' opportunities that month for tourists and it is titled 'Explore the wild!' (Yukon Department of Renewable Resources 2001b).

Gwich'in Human–Animal Relationships

A common theme with much of the communication between the various renewable resource departments and the public is that of 'the wild' or 'wilderness'. The idea of 'wild', 'wildlife' and 'wilderness' is problematic. As Tim Ingold (1993) points out with his analysis of globes and spheres, we should be careful of the indexical qualities of the terms people choose to express ideas about the environment. In much of Euro-Canadian thought 'wild' refers to situations where people have not interfered somehow or in some way with the equally constructed 'natural order of things.'[8] Wilderness, therefore, is something which is put aside and looked upon from what is considered to be a nonintrusive distance, or it is something which must be tamed or conquered through a process of domestication.[9] Keith Basso (1990) and Julie Cruikshank (1990; Cruikshank et al. 1990) through their descriptions of learning about landscapes, are clear that anthropologists should be prepared to listen to the Native teachers who are willing to talk to us about their landscapes so that we may come to a better understanding of how our informants or teachers, rather than ourselves, construct the World. By allowing his teacher to instruct him in a way that the teacher felt was appropriate, Basso was able to learn that among the Western Apache the landscape is not glossed by the opposition between people and nature; but rather it is catalogued through stories of the interactions between people and the land. Thus, it is not only that anthropologists must keep their ears open but also that they must allow a good deal of latitude within their methodologies so that their teachers will tell them what they know through a process that makes sense to them.

When Gwich'in I have worked with tell stories about the land and the animals, and when they refer in these stories to something as being 'wild', they mean that the animals are not acting in ways which are normal or they mean that the animals are not giving themselves to hunters anymore. Either way 'wild' is not considered to be positive because it signals a breakdown in the human–animal relationships. 'Wilderness' is a horrific concept for Gwich'in because it refers to situations where people can no longer enter into relationships with animals and starvation and social breakdown are imminent. Therefore, distancing one's self from the land through an idea like wilderness is not considered to be beneficial at all. Instead what is considered beneficial is that one comes to understand something about the proper interrelationships that occur on the land among animals, between animals and the land, and between animals and people on the land. It is when improper relationships are practised that wilderness is brought forth in narratives as a possible outcome of these actions (Wishart 1999).

On a couple of occasions I have heard non-First Nations people who live in the same area as the Gwich'in talk about how you almost never see Gwich'in people going out to just look at things. For example, one individual said, 'When the caribou come, you almost never see Gwich'in going out of their way in order to look down upon a valley full of caribou for the pure joy of seeing something like that.' While this is an overstatement, there is a good reason for a seeming lack of interest in what is considered to be wildlife viewing.[10] Gwich'in consider any action such as going out to stare at caribou as they go about living their lives as being a possible intrusion into their country. Any intrusion has the capacity to 'bother' these animals and cause them to become wild. When people and animals meet, it is considered to be a sign that 'God' has destined this moment so that animals can give themselves to the Gwich'in in order that they can survive. Therefore many other interactions between animals and people are seen as problematic and potentially dangerous or polluting. This respect for living spaces also occurs in reverse. When an animal wanders into a camp it becomes something that pollutes or puts people into danger. These are two different worlds which should be kept separate out of a respect which is deemed to be mutual.[11] People are given animals for their benefit and to intrude on animals' worlds by going out and bothering them for the purpose of fulfilling an aesthetic desire which conforms to a foreign logic could be a potentially dangerous habit.

Caribou, like all other animals, are considered to be sentient by Gwich'in. Probably the first thing one comes to realise when working with Gwich'in elders is that caribou, more than any other animal, are talked about as being animals who consciously choose their country. In other words, one comes upon many situations where caribou are talked about as avoiding places because they are being 'bothered' there by some sort of improper action or memory of improper actions. An example of the Gwich'in idea of caribou changing their ways in relation to things that are bothering them was told to me one day when an elder asked me to help him get water at James Creek. James Creek is in the mountains and it is about a forty-minute drive away from Fort McPherson, on the edge of the area where caribou often winter. While we drove along the highway further up into the mountains, he began to talk about when and where the caribou could usually be found. He pointed out areas in the mountains where he had successfully hunted in the past and told stories about the hunts. At one point along the road a long valley can be seen in the distance. At this point the elder pointed to the valley and said:

> you know those animals are funny sometimes, they know things and can change because of us. See that valley there, the one in the distance. They say there used to be lots of caribou there every year. You could go there and they would be just stacked up in there. One winter a guy went into that valley and disappeared, he died in there. Nobody ever found him, but since then caribou avoid that place. They know that something is wrong there.

While the elder would not explain why the disappearance of this man long ago should bother caribou, there is an implicit understanding that the two worlds of people and caribou are separate but are also inter-linked and mutually sustaining. Thus when a breakdown occurred in the way that people should act, the caribou responded in a negative way. People worry about any possible intrusions because caribou are so significant in relation to who they are. For instance, one of the ways that I have heard Tetlit Gwich'in describe themselves in relation to other people is that they are 'caribou eaters'. This self-description not only points out that their main source of food during the cold months of the year is caribou meat,[12] but it also makes a statement about the cultural significance of a group of animals which continue to come back to them because the people are living in an appropriate manner.

When wildlife biologists explain their actions by describing them as experiments (as in the case of the introduction of muskoxen), elders often react negatively. The whole idea of experimentation with animals, and particularly with anything that might influence caribou, is considered to be gambling with extremely high stakes. The stories about the experimentation with muskox are often told side by side with stories of people being cruel towards animals, teasing them or wasting them. The outcome of these stories is either comical or disastrous but it is always negative. On one occasion, after listening to a few elders tell these stories, the wife of the elder who shot the muskox – who is herself highly respected – pointed out to me that whenever any action is taken 'on the land', people must think of the grandchildren and their grandchildren. She then paused and said, 'when those people put the muskox there, they didn't think of how our people were going to eat'. However, the concern is not just that the introduction of the muskox happened as the result of an experiment with animals, but that muskox are also thought to negatively affect caribou.

Gwich'in Understandings of the Relationship between Caribou and Muskox

As noted previously, the Gwich'in are extremely concerned about the availability of caribou. Caribou are not taken for granted in any way and people talk about what may be affecting them. If the caribou did not come back to Gwich'in country, or if their numbers declined rapidly, the consequences could be disastrous. 'Wildness' in caribou is a factor which people must deal with. When caribou begin to act unpredictably, as was the case with the spring migration of 2000 discussed at the beginning of this paper, then people begin to look for factors that are bothering the caribou and causing them to act this way. And one of the concerns which people have is with the growing numbers of muskoxen[13] in the area where the caribou pass through on their migrations through the Richardson Mountains. Gwich'in believe that muskox are not meant to be in this area and they talk quite openly about how the 'biologists' put them there in Gwich'in country without asking anyone or being told by anyone to do it.

Muskoxen are considered to 'bother' caribou in two ways.[14] One elder described this process as the following: 'Those muskox they come and eat all the caribou food, they eat it right down to the ground. Also that muskox it pees all over the place and that pee really stinks. Caribou smell that pee from a long way and they go the other way.

They hate that smell and anywhere muskox go, the caribou don't go there anymore.'

This elder's testimony speaks to two beliefs about caribou. First is the idea that caribou are very selective about what they eat. Certain lichens and shrubs are the preferred food of caribou and they believe that muskox will eat the caribou out of the country. Second is the idea that caribou have an extremely good sense of smell and anything that smells out of the ordinary will cause the caribou to avoid that area.[15] Therefore people in Fort McPherson are very concerned about what they consider to be a muskox invasion of their country, and the idea that these animals were put there partly to lure tourists into situations where they will bother other animals runs counter to their own ideas about proper human–animal relationships.[16]

Discussion and Conclusion

What I have tried to do with the preceding example is to sketch out some of the differing views of the land between the Gwich'in that I have learned from and one particular level of renewable resource management in the area. In this part of Canada, the multiple layers of wildlife management organisations and their mandates is extremely complex and I would like to note that I am not trying to vilify these organisations – they are full of very dedicated people who believe in what they are doing and some of the things they do are beneficial to Gwich'in and other First Nations, and the people recognise this. However, what needs to be expressed is the confusion that surrounds the goals of these two interlinked camps. For instance, I have been working for the past four years on a project which has a sustainable forestry focus. After my first year staying with people on the land, I have focused on demonstrating how Gwich'in harvest wood in a manner which is not only sustainable but it is linked to other hunting and 'harvesting' practices and therefore has a great deal to do with what people consider to be their traditional manner of living on the land. Some of the biologists that I work with in pursuing this sustainable forestry project tend to point out that my work is not scientific and has little to do with trying to establish a system for managing timber resources. I cannot demonstrate which areas should be designated as cutting blocks and which should be protected. Indeed, everything I have learned from the people I stay with runs counter to this idea of dividing up the land. The Gwich'in have already divided their country through intricate kin networks and through human–animal relationships and are already highly selective in their choices when harvesting trees or indeed anything else. The biologists and management officials tend to argue that they want to protect the land and do not want the north of Canada to become like the southern regions, but the idea of mapping onto this area a new landscape of use and no-use areas would in my mind run counter to their goals.

The management organisations realise that, unlike in the southern parts of Canada where food, building supplies and fuel can be purchased with ease and at a reasonable cost, should the residents of Fort McPherson be forced not to hunt, trap, fish and cut trees, they would have to rely on goods imported from southern areas; goods which suffer both in quality and from inflation by the time they reach the community. Therefore the residents continue to rely upon sources of these goods from the land to fulfil their needs. However, what these organisations tend to ignore, or at least downplay, is that as important as fulfilling these tangible needs is to the survival of any people, an equally important aspect is the satisfaction which comes from harvesting these goods out on the land their way. Indeed these two aspects of Gwich'in practices cannot be separated, except in academic discourse. To the Gwich'in, hunting and staying on the land not only feeds them – it is also a key aspect of what makes them Gwich'in. This aspect is obvious to all people who maintain a hunting-gathering tradition but it is one which must be continually spelt out to many other audiences. Hunting not only enriches the people's lives by bringing food to the table, it is also a key element of who they consider themselves to be in regards to their history and to the relationships they have with the animals who are co-inhabitants of the landscape described as 'on the land'. So when the elder shot the muskox he did it not only for the meat; he did it also because, according to all the complex understandings and implications of what it means to live 'on the land' in his country, it was the proper thing to do.

Notes

1 For example, Feit (1988, 1991), Freeman (1985), Scott (1979, 1988), Scott and Feit (1992).

2 For example, Brody (1981), Brightman (1993), Fienup-Riordan (1994), Ridington (1990), Tanner (1979).

3 In a sense this paper reflects many aspects of the research I have been doing with the Tetlit Gwich'in for the last four years and as such I would like to acknowledge the following granting agencies who have provided financial support for the research: Social Science and Humanities Research Council of Canada; Jacobs Research Funds; Sus-

tainable Forestry Management Network Centre of Excellence; Faculty of Graduate Studies and Research, Research Fund, University of Alberta; Gwich'in Renewable Resource Board; Northern Science Training Program, DIAND.

4   The Dempster Highway is a narrow gravel road that winds its way through extremely rugged, although spectacular, mountainous terrain for about 750 km from Dawson City in the Yukon to Inuvik in the Northwest Territories.

5   If one looks at the life stories of many of the eldest members of the Tetlit Gwich'in one is struck by the fact that many of them were born on the land somewhere within the Yukon, and, according to many elders, this is due to the fact that many of the Tetlit Gwich'in used to spend a large portion of each year either hunting caribou in the mountains or trapping and fishing up the Peel in the Yukon.

6   Porcupine Caribou are the exception to these rules as any member of a First Nation represented on the Porcupine Caribou Management Board has the right to hunt Porcupine Caribou anywhere within their Canadian range.

7   The other two mechanisms are the Committee on the Status of Endangered Species in Canada (COSEWIC) and the Convention on International Trade in Endangered Species (CITES).

8   I realise of course that 'wilderness' has been the subject of many academic works in many different disciplines and is a complex, dynamic, and highly reflexive idea.

9   One should be careful to note that this has everything to do with cognition and little to do with fact. I would not like to argue that our wilderness parks are in any way 'natural', other than a perception of what that term means.

10 This is not to say that Gwich'in do not take great pleasure in seeing animals as they go about their lives on the land. For instance, I often hear elders talk about how in the spring when the geese come back the country looks 'just nice' because they are so happy the waterfowl have returned.

11 Domestic dogs have been the subject of a few interesting studies in relation to Native peoples' cosmologies. Among the Gwich'in, dogs are respected for their abilities such as pulling sleighs and as early warning systems for bear attacks but they are still considered to be dirty animals because they live within the confines of human society. For example, children's table scraps are never given to dogs, out of the belief that the children will become sick and never develop properly.

12 From late May to mid-October fish and, to a lesser extent, waterfowl are the most heavily consumed wild source of meat.

13 The Gwich'in from Fort McPherson think that the muskoxen seen from time to time have now established themselves within that area. From what I understand, biologists, on the other hand, seem to think that they just occasionally wander up there, as evidenced by how few reported sightings there have been. However, I have heard of many other sightings by Gwich'in than are recorded in the reports I have read. While these two bodies of knowledge may seem in contradiction, in actuality there may be some convergence of Gwich'in belief about the establishment of muskoxen and the scientific literature. Gwich'in have noted that only lone bulls used to be seen in the area and that now groups can be seen. Smith (1989b: 1096) notes that it is young bulls which 'pioneer' new habitat before acquiring harems.

14 Some of the information that Gwich'in use to establish a case against muskox comes from their contact with Inuvialuit from Sachs Harbour, where there is a muskox overpopulation problem and a depletion in caribou numbers.

15 A good number of Gwich'in hunting practices arise partially out of this observation.

16 While the politics of 'eco-tourism' is secondary to this paper's theme and argument, it is interesting to note that the Yukon Department of Renewable Resources (2001b) has declared that, 'A hunting closure will be announced for a one week period as the [Porcupine Caribou] herd reaches the [Dempster Highway], giving an excellent opportunity for wildlife viewing. Keep your eyes open in the local media for this time announced. Look for viewing tours to welcome the caribou's return.' This closure was originally introduced (at the insistence of elders) because in respect for Gwich'in tradition when the caribou first arrive people should not bother them, so that the leaders of the herd will remember that it is good country to pass through or to stay in for the winter; i.e., so that they will not act as wild as they have been.

6

'We did not want the muskox to increase': Inuvialuit Knowledge about Muskox and Caribou Populations on Banks Island, Canada

Murielle Nagy

Introduction

This chapter compares the knowledge of Inuvialuit and that of biologists about how muskoxen and caribou relate to one another. I focus upon one contrasting assumption: the assumption on the part of biologists that the two species are discrete, as opposed to that of Inuvialuit of the patterns of interrelation and avoidance between the two. This debate takes place in the context of one of the most dramatic and puzzling environmental shifts (some would say disaster) in the circumpolar North: the abandonment of Banks Island by Peary caribou in the years following a massive increase of muskoxen. I conclude in this chapter that indigenous knowledge provides us not only with hints as to the rich behaviour of Arctic mammals, but also with wisdom on how to arrange wildlife so as to suit local needs. In this case, Inuvialuit speak in polite admonishment of the policies of wildlife agencies, which made muskox increase without taking into account the Inuvialuit knowledge that this would only lead to emptying the land of caribou. In this chapter the application of knowledge of animals implies certain choices of what the landscape will look like, which is in the end a political issue.

Muskoxen are very much part of the imagery of Arctic landscapes. Authors writing on the muskox (e.g., Gessain 1981; Lopez 1986) have a definite admiration for an animal that, having originated in the Old World and survived the ice ages in North America, was considered an endangered species in the early twentieth century. Muskoxen were overhunted during the nineteenth century in the Canadian eastern sub-Arctic to provide hides to the Hudson's Bay Company. To ease pressure on muskox populations in Canada, the animal was first banned from hunting and then protected through conservation laws. Biologists became involved in the reintroduction of muskoxen in areas where they had disappeared, and even in their introduction in areas where they had never lived (Lent 1999). The muskox was, and to a certain extent still is, seen by some as the saviour of Aboriginal people of northern areas where other subsistence resources, such as caribou and fur-bearers, have been fluctuating. Arctic explorer Vilhjalmur Stefansson (1924) even dreamed that muskoxen were to be domesticated and would provide meat and wool for the Inuit and the world market. His belief that muskoxen were related to European Highland cattle was consistent with his notion of the 'friendly Arctic', a land that was productive and viable rather than barren and desolate (Stefansson 1921)

Due largely to the Hudson's Bay Company exploitation of muskox hides from the mainland during the second half of the nineteenth century, it was assumed by the mid-1910s that the muskox population in Canada was on the verge of extinction and the federal government therefore forbade their hunting in 1917 (Barr 1991; Lent 1999). Geographer Barr (1991: 42–43) links this conservation legislation to the pressure exerted on the Canadian government by scientists such as biologist Gordon Hewitt, who recommended that the hunting of muskoxen be prohibited on Victoria, Banks and Melville Islands. Hewitt wanted these islands to become permanent reserves for muskoxen. Historian Dick (2001) associates this legislation with the fact that in the High Arctic, the expeditions of American explorers such as Robert Peary (1898–1902, 1905–6, 1908–9) and Donald MacMillan (1913–17) had hunted hundreds of muskoxen on Ellesmere Island. However, it is also clear to Dick (2001: 273–77) that the Canadian government was mostly asserting its sovereignty over the Arctic islands, and that such a legislation would have to be enforced by sending police patrols and establishing police detachments. Indeed, Canada feared that Denmark claimed Ellesmere Island because the Inughuit of northwest Greenland travelled to hunt there regularly. There was thus a political agenda behind a law aimed at protecting muskoxen. One might even wonder if muskoxen of the Canadian Arctic were indeed an endangered species.

In the case of Banks Island, the westernmost island in the Canadian Arctic Archipelago, located in the Northwest Territories and covering about 70,000 km², its muskox population was considered an endangered species in the first half of the twentieth century. However, since the 1970s, their numbers have undergone a drastic increase due in part to the fact that the Inuvialuit, the Inuit group that live in the western Canadian Arctic, were forbidden to hunt muskox on the island. The result, anticipated by the Inuvialuit in the 1960s, was that the caribou population almost disappeared on the island. The Inuvialuit would have preferred to keep the muskox population at a low level, since they knew from oral tradition that a high population would drive away the caribou, which they prefer for their subsistence.

In this chapter, the information and quotes from the Inuvialuit, mainly elders, are English translations and transcriptions of archival tapes and interviews collected in 1996 during the Aulavik Oral History Project, a study that was undertaken by the Inuvialuit Social Development Program (Nagy 1999).[1] Throughout the text, I contrast Inuvialuit views with those of biologists who have been working on muskoxen and caribou. Three key issues are discussed. The first is about the conservation philosophy and policies that federal and territorial governments, as well as wildlife scientists, imposed on the Inuvialuit up to the early 1990s. The second demonstrates that although biologists did not integrate Inuvialuit environmental knowledge to their studies, over the years they changed some of their views on muskoxen and caribou, and their data now parallel local models on those animals. The third issue is about the choice of species. The governments and wildlife managers chose to let the muskox population increase while the Inuvialuit wanted to keep it at a low level for fear that they would drive the caribou away (which, in the event, happened). This is a clear case of government policies aimed at the protection of one species–muskox–without much regard to the concerns of the Inuvialuit who should have been allowed to control the muskox population in order to keep a sustainable population of caribou for their subsistence. Such wildlife management was probably practised at times by past human occupants of Banks Island.

Banks Island Muskoxen in the Past

Archaeological sites on Banks Island have shown that muskoxen have been hunted there for at least the last 3,400 years (Müller-Beck 1977). But when Stefansson explored the island in the mid-1910s, there were only a few muskoxen. Stefansson (1921: 241) was convinced that the Inuinnait (also known as Copper Inuit in the literature) had overhunted muskoxen during their summer visits to the northern part of the island after 1853, while scavenging metal and wood from the shipwreck of the HMS *Investigator*. This heavily copper-sheathed vessel, guided by Captain Robert M'Clure, was part of a British search attempt for the lost Franklin expedition. M'Clure abandoned the vessel after it became trapped in heavy ice in the spring of 1853. However, some of the numerous archaeological sites along the Thomsen River pre-date the visit of M'Clure's crew at Mercy Bay, which weakens Stefansson's conclusion (Wilkinson and Shank 1975; Toews 1998:141–44). This said, muskoxen remains predominate many assemblages from archaeological sites of Banks Island (e.g., Will 1985; Woollett 1991). From this evidence, muskoxen might have been the major ungulate exploited on the island either because caribou were few or absent, or in order to control muskox population, a subject that will be discussed later in this chapter.

In 1915, when Stefansson was on Banks Island, there were indeed only a few (if any at all), muskoxen on the island, as recalled by Susie Tiktalik who was a young teenager at the time: 'it was at that time when there was no muskox there too. We did not see any muskox tracks that summer when my parents were walking the land. There was no muskox for a long time' (ST:N92-253-213B: 5). The same year Stefansson was on Banks Island, he asked Susie Tiktalik's father, who was named Kullak, if muskox had been extinguished by hunters. Kullak answered that muskox had moved away (Stefansson 1921: 370). Asked the same question in 1996, David Nasogaluak answered, 'I don't think so. Not enough people to hunt them and when Nature put ice on the ground…, they starve right away' (DN:Aulavik-47A: 5). This explanation was also favoured by Frank Kuptana, who said that 'long ago, Banksland had muskox. Then land froze up, iced land. The muskox had nothing else for feeding. The muskox became less and less, just like the caribou [nowadays]' (FK:Aulavik-27A: 3). Natuksiaq (also known as Billy Banksland), who was the guide of Stefansson, made similar comments in 1939: 'All the musk-oxen are gone. They died long ago when the rains came and froze on the land, covering the grass with a mantle of ice and leaving nothing for them to eat' (de Coccola and King 1986: 334). Such weather-related depletion was also experienced by Susie Tiktalik, as told by Sam Lennie: 'they had big rainstorm, and muskox … died in Banksland, everything that lived of the ground, [like] caribou, everything. [Tiktalik] said in springtime, Banksland was so stinking [that] they all went back to Victoria land' (SL:Aulavik-13A: 2–3).

Michael Amos recalled that Susie Tiktalik often said that three years after people killed off the muskox, the caribou started coming back: 'They never saw any more muskox, they cleaned them right out that time. The muskox, they had been killing them all that time because there was going to be no more caribou' (MA:Aulavik-78A: 3). Sarah Kuptana also heard from her husband William Kuptana that 'long ago they finished the muskox by doing that. The Qangmalit [eastern Arctic people] would surround big herds and kill them. Then, there was no more muskox, but the herds grew again' (SK:Aulavik-60A: 5).

Lopez (1986: 43–46) has pointed out that in the nineteenth century Inuit hunters might have killed off the muskoxen on Banks Island. Lopez gives examples of animals being hunted to the point of extinction elsewhere in the world but adds that we do not know the motivations of the hunters, nor if they understood the consequences of their acts. But why deny Inuit hunters from the past the ability to decide how to manage animal populations? I think that in general we are looking at overhunting as a bad practice and we do not even try to find an explanation for it. As is well illustrated by Krech (1999), models on aboriginal people's exploitation of animals portray them as either naive ecologists not interfering with Nature or careless hunters extinguishing fauna (Martin 1973). What about a middle ground, where aboriginal people have preferences for some species and manage animal populations in consequence?

Although it is possible that hunters on Banks Island exploited muskoxen because they were the major resource on the island, it seems likely that in the nineteenth century, and maybe even before, people tried to keep some control over the muskox population to protect the caribou. In other words, before the arrival of Europeans, traditional management of muskox and caribou might have been to maintain low muskox levels to make sure that caribou population would not decrease. Indeed, this would explain the reason why 'human hunting was a major influence on the numbers and distribution of muskoxen around the circumpolar north prior to the introduction of firearms' (Lent 1999: 216).

As we can see, it is not clear whether the absence of muskox on Banks Island in the 1910s was due to migration, starvation through a natural event, or intense hunting in order to have caribou on the island. However, one point that was often stated by Inuvialuit elders was the fact that a high muskox population is correlated with a low caribou population, or even no caribou at all. Regarding Banks Island, Inuvialuit mentioned that 'There was not much caribou long ago. They hunted down muskox for food and made dry meat' (William Kuptana in WK:N92-253-367A: 1–2). 'Long ago, there was hardly any caribou in Ikaahuk (Banks Island) but there was lots of muskox' (Sam Oliktoak in SO:Aulavik-35A: 4). 'There used to be a lot of muskox long ago and they would eat only muskox through the winter and summer' (Mark Emerak in ME:Aulavik-74B: 4). And, as Helen Kalvak put it, 'Long ago, in early May when [muskox] are just born, people used to hunt young muskox with no weapons. They used to hunt them for clothing because long ago there were hardly any caribou on the land' (HK:N92-091: 10–31).

Increase of Muskox

In the late 1950s, muskoxen made a comeback to Banks Island, but it was only in 1970 that an estimated 1,567 muskoxen were counted for the northern part of the island (Kevan 1974 in Lent 1999: 194). However, biologists and wildlife managers were not ready to accept such a number since observers in previous decades had reported only scattered muskoxen (Lent 1999: 194). But as Lent (1999: 195) noted: 'if a population of 100 animals had existed in 1950 it could have grown to the estimated 1970 level under ideal conditions with no hunting pressure but without immigration'. In 1972, 3,800 muskoxen were estimated for the whole island (Urquhart 1973 in Larter and Nagy 2001b: 395). According to Susie Tiktalik, what caused the increase of muskox was migration to the island: 'when the muskox population wanted to go up, it went up. They must have crossed from somewhere' (ST:N92-253-213B: 5).

In his article on Banks Island, Douglas (1964: 710) mentioned that the people of Banks Island linked the disappearance of the wolf to the resurgence of the muskox. Peter Esau also thinks that the near absence of predators has influenced the growth of muskox on the island: 'there's not much wolves now too. Long ago there used to be a lot of wolves, but they really poisoned them for the trappers. People did their trapping by dog team but the wolves ate too much white foxes from their traps. That is why they were killing them, the game wardens ... came around here and they were poisoning them' (PE:Aulavik-24B: 1). Nowadays there are wolves again on Banks Island and some are harvested by trappers (see Nagy et al. 1996: 219).

In the 1960s, Susie Tiktalik advised people that muskox population should be kept low in order to have caribou on Banks Island, as recalled by Sam Lennie: 'Susie Tiktalik told hunters and trappers, "if [there is] anyway you fellows could destroy all the muskox, do it". She said [that] when she was a very young girl ... there were a lot of caribou but muskox migrated through there and chased all the caribou away' (SL:Aulavik-13A: 2–3). Many people, like Sam Oliktoak, think that the government should not have kept the severe restrictions on the hunting of muskoxen: 'long ago the white man said we couldn't kill any muskox. Because the muskox weren't being killed anymore, now there's really lots around. Even in the summer, the people don't walk the land anymore, that's why. They don't hunt those animals anymore' (SO:Aulavik-35B: 9). People from Sachs Harbour voiced their concerns, as remembered by Agnes Carpenter:

I know the hunters and trappers had written out to the Game people and to Ottawa. That was ... right in the early '60s, I know we started writing about muskox. That was a species at that time that was prevented from being killed. It was outlawed to kill muskox. The government was protecting them so they could make a comeback. But the people were against it, that's what we wrote to tell the government, *that we did not want the muskox to increase* because from past experience, past history of what the elders knew, ... muskox were competing for the same food that the caribou were eating, ... they were going to completely wipe out the caribou herds. There's been meetings galore, there's been writings about it, there's been tapes on it, there's been recordings on it, but the government would not budge. And by the time they open [the hunt], the government kept having surveyors going onto the northern part of the island, the muskox were breeding like nothing! In ... just a few years, it went from almost a few, just a few hundreds maybe not even that, up to 10,000. (AC:Aulavik-12A: 7) (my emphasis)

The muskox population on Banks Island has steadily increased since the 1960s. It was estimated at 29,168 (±2,104) in 1985, and reached a peak of 64,608 (±2,009) in 1994 (Larter and Nagy 2001b: 394). Although the population has since decreased, in 1998 it was still high at 45,833 (±1,938) and the world's largest indigenous population of muskoxen (ibid.: 395). The quick growth of muskox population of Banks Island within the past thirty years and, in the 1960s, the reluctance of the government to change its law which forbade the hunting of muskox, show that although people warned that caribou might disappear from the island due to muskox expansion, oral history and traditional knowledge were not taken seriously. The view was very much that expressed by biologist Lent (1999: 272): 'even if traditional ecological knowledge were being adequately maintained, passed on and applied to resources management, changes in northern ecosystems are occurring that are beyond the scope of that knowledge'. Wildlife management was not regulated for the interests of the Inuvialuit but with the goal of bringing back a substantial population of muskoxen on Banks Island without concerns for the long-term impact on caribou population. Yet, as Lent (1999: 193) suggested: 'an alternate management approach would involve holding the population to less than the level of maximum yield and manipulating it to reduce the population growth rate rather than achieving maximum growth'.

The ban on muskox hunting was finally lifted in 1971, with an annual quota of seven muskoxen that was raised to 150 in 1978 (Gunn et al. 1991: 189). Such low quotas did not allow hunters to control the rapid growth of the muskox population, as explained by Andy Carpenter and Peter Esau:

Andy Carpenter: Around the 60s they'd seen about 200 [muskox] [south of Mercy Bay]. Then in the 70s that's when they started really increasing … . There was about 3,000 to 4,000 muskox. Lots of them, and then they started coming this way. We started seeing some around [Sachs Harbour].

We tried to tell when we saw the increase coming, we tried to tell the government so they would start taking some of them. And they gave us so small quota, when they first open the quota that was in 1971. They opened it for 15 muskox and they had to be hunted [in what is now the Aulavik National Park in the northern part of the island]. … You had to get them on other side of Bernard River. Nothing close. Now, you can't travel all the way up here to get one muskox and that's all you could get, one muskox. … they couldn't take the quota anyway because this was too far. (AC:Aulavik-20B: 9–10)

Peter Esau: When they first opened the hunt, … we had permission to get eight muskox. Apian and I we were the first ones … to hunt muskox. We got two. One each. We went by skidoo. … They had that quota till they started to come [south]. They tried to manage them; they were not managing right. That's why they over populated, we could have controlled them right from the beginning, you know. We could get hunters and everything up here, but they were [called] 'endangered species'. They called them that for so long and, they never wrote me for sport hunting. Now there is too many. You know, the white people they over do it sometimes (PE:Aulavik-24A: 12–13).

At one time we got 11 muskox, there were five cows altogether … . They all had young ones inside of them. … They are always born one right after the other. That's the way I found out myself, the white men talked, [but] I really proved it myself.

We talked [with biologists but] they still don't believe in it now yet. When we started doing the killing, we saw a cow here that was a two year old, we saw it, it had young one inside of her. It was gonna have a young. She was age of two … yet they were saying that they will not have youngs till they are four years old first. But it's so true, we saw it ourselves. When they are two years old, they have a lot of youngs. They really increased. (PE:Aulavik-24B: 1)

In the 1960s the prevailing views among biologists were that female muskoxen did not produce their first calf before age four, and that birth of calves in alternate years was the rule (see Lent 1999: 158). Lent (1999: 177) seems to have been one of the first biologists to record female muskoxen that reached sexual maturity as two-year-olds in the 1970s. In 1982 and 1983, calving by two-year-olds was recorded in Banks Island but the sample sizes were considered too small to be representative (Rowell 1989 in Gunn et al. 1991: 190). However, in the early 1990s, information collected during commercial harvests revealed about 30 percent pregnancy by two-year-old females (Larter and Nagy 2001b: 396). Hence, before 1993, biologists working on Banks Island generally defined adult females as three years old or older. But after 1993, surveys included two-year-old female muskoxen as adults (Larter and Nagy 1999: 3), thus corroborating the above–mentioned observations of Peter Esau.

Decline of Caribou

The Peary caribou population of Banks Island seems to have been on the increase from the early 1910s to the 1970s. It was estimated at 2,000 to 3,000 in 1914 (Stefansson 1921), 4,000 in 1952–53 (Manning and Macpherson 1958: 65) and 12,000 in 1972 (Urquhart 1973 in Larter and Nagy 1997: 9). One should note, however, that before the 1980s, the survey methods and the population calculations were less sophisticated, hence the estimates might have been lower than the actual number of caribou. In the 1950s and 1960s, the average caribou take was 14 per hunter (Usher 1971: 71). However, the caribou population on the island declined to 6,970 (±1,133) in 1982; 897 (±151) in 1991; and 1,005 (±133) in 1992 (Nagy et al. 1996: 213). In reaction to that decrease, the Sachs Harbour Hunters and Trappers Committee established a quota of 150 caribou in 1990; 30 males in 1991; and 36 males in 1992, to allow each family to harvest one caribou (ibid.). The caribou decrease continued and, in 1994, it was down to 709 (±128) (Larter and Nagy 1997: 9) and to 436 (±71) in 1998 (Larter and Nagy 2000: 661). Agnes Carpenter and David Nasogaluak made the following comments on the decline of caribou:

> *Agnes Carpenter:* The caribou herds kept declining all the time. Then gradually the muskox moved from the northern part of the island. That's [where] they were breeding, on the northern part of the island. They gradually came down. They kept pushing the caribou herds down and finally in the end we had hardly any caribou left. The caribou used to migrate up to the northern part of the island during the summer months, and they migrated back down towards the fall.
>
> In the end we had nothing coming back. Hardly nothing coming back and there, caribou were sort of going, staying along the coast line. ... There was hardly anything on the inland. ... It'll take years and years and years for the caribou to come back. (AC:Aulavik-12A: 7–8)

David Nasogaluak: The first time I did survey work was with Dr. Doug Urquhart from Fish and Wildlife, he was surveying this island for caribou. ... We surveyed caribou, there was 14,000 caribou one year but gradually going down. [Moving] east, nobody killed them off or anything. They gradually moved out from the muskox. ... They have to move to the daylight, the daylight comes from east. That's why they always migrate towards the east. In the spring time, the caribou are heading north to the daylight. In the fall, they start migrating when daylight is coming, finishing off in the south side of us. That's why the caribou's migration, nobody know them much I think. I myself know that from studying caribou and reindeer. (DN:Aulavik-47A: 4)

David Nasogaluak's comments are a good example of how Inuvialuit environmental knowledge is affected by scientific knowledge. As Peter Usher (2000: 185) stated: 'field science programs have been employing aboriginal Northerners since at least the 1960s, including some who are elders today. They are aware of what scientists actually do and find out, and even if they do not agree, they have considered scientific knowledge critically against their own.' Yet local knowledge has not been readily accepted by scientists. A case in point is the Inuvialuit affirmation that when muskox became numerous on Banks Island, they started to compete for the same food niche as the caribou, as expressed by Frank Carpenter, Andy Carpenter and Geddes Wolki:

Frank Carpenter: When they first started [coming], they were not really bothering [the caribou] because muskox feed on the valleys, you know, where there's grass, and the caribou, on the hills. But we started to notice that they started going on the hills later on. They never used to. They just stayed in the flats, where the grass is. [Near] Raddi Lake, [the muskox] just go right out on a point and go up, ... muskox were eating everything to the ground. And the caribou, they're waiting just for top. (FC:Aulavik-14A: 9)

Andy Carpenter: [The caribou] started moving out. There were lots of them. Caribou have moved out around Victoria Island. ... The caribou started going down, next thing you find out, there's no more caribou in this area. We keep telling that there's competition between caribou and muskox. [Biologists] still didn't believe us. (AC:Aulavik-20B: 11)

Geddes Wolki: [Muskox] eat so much, maybe they take all the food and let [the caribou] get short of food, maybe. You know the big muskox can eat three times more than one caribou, or even four times as much. [They have] big guts. (GW:Aulavik-26B: 4)

That caribou competed poorly with muskoxen for forage had been stated by Inuvialuit for some time (e.g., Lopez 1986: 43) but at first it was not taken into account by biologists since it was not supported by their data. Most were of the opinion, expressed by Lent (1999: 195), that 'a cause-and-effect relationship between the increasing numbers of muskoxen and the declining population of caribou *has not been clearly demonstrated*' (my emphasis). In the case of Banks Island, Wilkinson et al. (1976) researched the feeding habits and range of muskoxen and caribou in 1973 but found no evidence of forage competition. However, their study was somewhat limited since it was done only during the summer. The possible role of muskoxen in the decline of caribou is also a subject of debate in Western Greenland (Lent 1999: 189). There, researchers have found that in winter the diet of muskox overlaps that of caribou (Forchhammer 1995). Recent research by Larter and Nagy (1997) has demonstrated that in Banks Island both muskoxen and caribou feed on willow and sedge, although the diet of caribou is more diverse. Their results have also shown that in Banks Island caribou and muskoxen have considerable similarity in their annual diets, especially in areas of high muskox density, and that it may increase during winters with elevated snow depth and density. Although they wrote that 'data cannot disprove or prove that forage competition has occurred or is occurring between muskoxen and caribou', they admitted that 'the potential impact of muskoxen on the caribou winter range and availability of willows may well be a factor limiting the recovery of the Peary caribou population' (Larter and Nagy 1997: 15).

Climate change has been reported by Inuvialuit of Banks Island (see Riedlinger and Berkes 2001) and the disappearance of some caribou on Banks Island might also be linked to climate conditions, as discussed below by Frank Carpenter and Peter Esau:

Frank Carpenter: ... In the '70s I guess, that's when they really started noticing it, muskox taking over. But [regarding] caribou, sometimes ... in the fall, we get freeze-up on the whole island. Then, before the snow is really deep, we get our mild weather and rain. Then it's cold enough for the rain to freeze on top the snow and that's when the caribou try to leave the island, even go out into the ocean. 'Cause they were eating mostly ice.

We were still here when one year it happened. When dogs started seeing the caribou, they'd be running. Nothing wrong with them but they just stop and start kicking. They have too much water in their stomach, their heads are spinning. So a lot of big bulls died off by spring ... . There was even one year, that worst year that time, the cows didn't have any calves, they didn't. That hit them just before the rutting season. (FC:Aulavik 14A: 10)

Peter Esau: I don't think [the muskox] really pushed the caribou away. Like right now the caribou are just dying, now. ... in the fall time, ... when the weather is not good, the ones that are born, they just freeze when the weather is not good. (PE:Aulavik-24A: 12)

Using the results of Wilkinson et al. (1976), Gunn et al. (1991: 192) dismissed forage competition and linked the disappearance of caribou on Banks Island to changing climate conditions associated to earlier spring snow disappearance, warmer winters that are snowier (hence more difficult for forage) and with higher incidence of freezing rain. Although annual die-offs of sixty to 300 caribou occurred during the winters of 1987–88, 1988–89, and 1990–91 when freezing rains occurred (Nagy et al. 1996: 213), Larter and Nagy (2000: 661; 2001a: 127) concluded that the drop in number of Banks Island caribou in 1994 and in 1998 happened despite high calf production, high over-winter survival rates of calves and less severe winter snow conditions. Thus, severe winter weather might not be the major cause of caribou decline.

Some Inuvialuit think that caribou do not like the strong smell of muskox and prefer to be away from them. Accordingly, caribou have moved out of the island to avoid muskox (e.g., Agnes White in AW:Aulavik-69A: 3). As remarked by David Nasogaluak: 'that Old Lady Tiktalik used to say that the smell of muskox, the caribou don't like it' (DN:Aulavik-47A: 4). Lent (1999: 202) noted that reindeer herders in Alaska believed that 'caribou and reindeer will avoid muskoxen, moving away when muskoxen enter their vicinity' but added, 'there is no quantitative evidence to support this contention, nor has a controlled study been undertaken' – hence expressing some of the distrust wildlife scientists might have towards local knowledge.

At least one person, John Lucas Sr., mentioned that the decline of caribou on the island might be due to their natural population growth cycle, something biologists have been able to quantify in other parts of the Arctic (e.g., Vibe 1967: 163), and that they would eventually increase again: 'the people figure that muskox are probably chasing the caribou away. But I don't think it's that way. I think it's probably what they call a 30 years cycle that they have the caribou. 'Cause, eventually I think they'd probably gonna come back. Maybe it's just a downfall' (JL:Aulavik-23A: 4).

What to do with the Muskox?

Despite the large size of Banks Island, there is a risk that a high muskox population will eventually damage the environment, and dampen recovery of the caribou. As Sam Lennie remarked: '[Muskox are] just like sheep, when they eat, they eat right to the roots and they don't leave anything' (SL:Aulavik-13A: 2). The environment of Banks Island might have already been affected since Larter and Nagy (2001b) recently reported that muskoxen population dynamics on the island from 1986 to 1999 was a density-dependent response, possibly related to food, like the availability of sedge in the summer diet of calves.

Most Inuvialuit agree that the number of muskoxen on Banks Island should be controlled. Although biologists Larter and Nagy (2001b: 395) qualified the muskox increase as 'spectacular', most biologists do not see the high numbers of muskox as a problem, but rather as a natural process (e.g., Lent 1999: 193; also A. Gunn's comments in Struzik 1995: 55). However, this increase was largely due to the hunting ban imposed on the Inuvialuit. Lent (1999: 192) even wrote that in areas where caribou have decreased, muskoxen could be used as an alternate meat source to ease harvest pressure on caribou. But this is a rather twisted suggestion since the problem started when the Inuvialuit were not allowed to manage the increase of muskoxen on Banks Island in the 1960s. Why should the Inuvialuit shift hunting preferences from caribou to muskox? Some Inuvialuit think that a too large population of muskox cannot keep healthy and will eventually die off. As Peter Esau put it, 'Tiktalik ... said there used to be a lots of muskox long ago. ... She said they die when they get too many of them, they start dying off' (PE:Aulavik-24A: 13), while David Nasogaluak said: 'They'll die again just like long ago. Right here, at Kellett River, there are muskox horns piled up all over. If it iced up, you'll lose them all in one season. Muskox are not travellers, they'll be weaker and weaker before they reach anywhere. They don't migrate. They say when there's too many in one place, they die off. They will do that' (DN:Aulavik-47A: 5).

Inuvialuit efforts to obtain a higher muskoxen quota to reduce their population, and hence hope that the caribou population would increase, were finally heard in the late 1970s. But what brought changes in the muskoxen quota was the fact that the Sachs Harbour's Hunters and Trappers Association 'widened its concern from the effect of muskoxen on caribou to *expressing fear* of a drastic crash on the muskox population' (Gunn et al. 1991: 189) (my emphasis). It is doubtful whether the Inuvialuit were really worried by the possibility of a muskox crash on Banks Island. Instead, they most likely realised that the only way to get a higher quota was to speak the same language that biologists and wildlife managers had used all along: that of 'protecting' the muskox. In 1981, a commercial harvest programme was initiated, and during the first ten years, the average commercial take was 124 muskoxen per year (range 0 to 269) (Nagy et al. 1996: 213).

With the signing of their land claim in 1984, the Inuvialuit became co-managers of the wildlife on their territory. The wildlife harvesting and management principles of the 1984 Inuvialuit Final Agreement (IFA) are to 'protect and preserve the Arctic wildlife, environment and biological productivity through the application of conservation principles and practices' (DIAND 1984: section 14.1). Under the Final Agreement, the Wildlife Management Advisory Council (NWT) was created and is the body that determines the 'total allowable harvest for game according to conservation criteria' (ibid.: section 14.36a). Inuvialuit Hunters and Trappers Committees were established for each community, and so was the Inuvialuit Game Council. The latter includes one member of each Inuvialuit Hunters and Trappers Committee and 'advise[s] the appropriate governments... on policy, legislation, regulation and administration respecting wildlife, conservation, research, management and enforcement' (ibid.: section 14.74b). Although there is a recognition in the Inuvialuit Final Agreement that the 'relevant knowledge and experience of both the Inuvialuit and the scientific communities should be employed in order to achieve *conservation*' (ibid.: section 14.5, my emphasis), Inuvialuit environmental knowledge has yet to be incorporated into research on caribou and muskox of Banks Island. This said, Inuvialuit have been involved in harvest research on different species through the Inuvialuit Harvest Study and the studies carried out by biologists.

The political power and co-manager status that the Inuvialuit gained through their 1984 land claim certainly helped increase the muskox quota. In 1986, the annual take for the Northwest Territories 'was up to about 800 muskoxen, of which half or more came from Banks Island' (Lent 1999: 196). By 1988, the total annual quota of muskoxen in the Northwest Territories had risen to 2,537, which included 2,000 for Banks Island (ibid.). Such a rise in quotas for Banks Island demonstrates that the Wildlife Management Advisory Council (NWT) started to realise that the muskox population needed to be reduced, or at least controlled. Muskoxen have been also hunted by the Inuvialuit for subsistence, but since 1988 the annual harvest has been fewer than 300 animals (Larter and Nagy 2001b: 216). In addition, Inuvialuit have conducted guided sport hunts of muskoxen for nonresident hunters (Nagy et al. 1996: 213).

With conservation and biological reproductivity being part of the philosophy underlying the wildlife harvesting and management principles of the Inuvialuit Final Agreement, it must have been somewhat difficult for the Inuvialuit Game Council and the Sachs Harbour Inuvialuit Hunters and Trappers Committee to convince the Wildlife Management Advisory Council (NWT) that the muskox quota had to be increased drastically. In fact, it was only in 1991 that the allowed quota was raised to 5,000 (Nagy et al. 1996: 213). The year before, the first large-scale commercial harvest had taken place with a total of 494 muskoxen (ibid.). In 1991, 1992 and 1993, the muskoxen commercial harvests were respectively 2,031, 1,798, and 738 (Larter and Nagy 1999: 15). In 1994, the total quota for Banks Island was up to 10,000 with no restriction as to age or sex (A. Gunn, pers. comm. in Lent 1999: 195). In 1997, 1,300 muskoxen were commercially harvested (Larter and Nagy 1999: 15). All commercial harvests were conducted from the southern part of the island (Larter and Nagy 2001b: 396), which is also where Sachs Harbour, the only settlement on the island, is located.

Lent (1999: 211) seems to worry that the 'quota for Banks Island alone, *if actually harvested*, would exceed the number thought to have been killed in all of Canada in any year at the peak of the nineteenth-century exploitation' (my emphasis). Speaking of the Thomsen River, he has little faith in the new quotas: 'humans may have played a key role in the former loss of muskoxen from this area in northern Banks Island. It remains to be seen whether a high population level will persist under a regime of regulated hunting' (Lent 1999: 277). But as Larter and Nagy (1999: 15) indicated, commercial harvests have had no major effect on the muskoxen population size. Furthermore, due to the small size of the community of Sachs Harbour, whose total population is 125 (Larter and Nagy 2001b: 395), and possibly the lack of processing facilities and of a market for such a high number of muskox products, the annual 10,000 quota has never been achieved on Banks Island.

How to control the muskox population is still an issue of concern for the Inuvialuit, yet not an easy one to resolve despite various recommendations on how many should be killed. As Robert Kuptana, who was part of many meetings on the subject, said: 'Those muskox have been growing in population very quickly. If they could get at least 10,000 a year, even that would not be enough because they just keep growing in numbers. Some of them, they could bring them to places where there are few muskox around' (RK:Aulavik-42B: 5). Based on knowledge of previous commercial muskox harvests, many Inuvialuit elders express their doubts as to how current plans to harvest the animals are going to work in practice. However, Lawrence Amos, an Inuvialuk from the younger generation, envisions much more than selling muskox meat. His solution is also about by-products, such as wool, and jobs for the Inuvialuit:

> The Inuvialuit of Sachs and even the whole region should just get together and start up some kind of a tannery. A place where you could have like wool on the side. 'Cause it's all gonna pay for itself. And it makes the people want to rely for themselves. Like you know the more effort they put into it, the more results they see for themselves. They don't have to depend on the government. (LA:Aulavik-15A: 10–11)

Conclusion

Archaeological sites on Banks Island have shown that muskoxen have been hunted there for many centuries. By 1917, the Canadian government assumed that the muskoxen were near extinction in the north and their hunting was forbidden. In the late 1950s, muskoxen made a comeback to Banks Island but the territorial and federal governments kept severe restrictions on their hunting despite warnings from the Inuvialuit that their population would overgrow and caribou would disappear. Indeed, according to oral tradition, before the arrival of Europeans, traditional wildlife management was to maintain low muskox levels to make sure that the caribou population would not decrease.

This chapter has presented and considered thoughts of Inuvialuit elders on muskox increase and caribou decrease on Banks Island. According to many of them, muskoxen are directly linked to the caribou decrease because they are competing for the same food, and are much more successful at it, an idea that was originally refuted by biologists but that is now accepted. Furthermore, many Inuvialuit interviewed said that the government should have allowed them to hunt muskoxen in the 1960s in order to restrict their rapid growth, and hence avoid the predictable decrease of caribou. The Inuvialuit were allowed to co-manage the muskox only after their land claim was signed in 1984 but it took another ten years for them to obtain a quota that could have reduced the muskox population. But in the mid-1990s, the muskox population had reached an estimated 64,600, while the caribou were reduced to less than 700. Hence, by that time, the muskox population had increased so much that it became unmanageable.

The quick growth of the muskox population of Banks Island within the past thirty years and the reluctance of the federal and territorial governments to change at first a conservation law which forbade the hunting of muskoxen, and then later their muskox management strategy, show that although the Inuvialuit warned that caribou might disappear from the island due to muskox expansion, oral history and traditional knowledge were not taken seriously. Nor was the preference of the Inuvialuit for caribou meat. In the mind of the Inuvialuit, for the government officials and wildlife managers, the wellbeing of muskoxen was more important than that of the people who lived on Banks Island.

Acknowledgements

The Aulavik Oral History Project was administrated by the Inuvialuit Social Development Program (Inuvik). Funding and logistical support was provided by Parks Canada, the Language Enhancement Program (GNWT), the Polar Continental Shelf Programme and the Inuvik Research Center. Sincere thanks to all the Inuvialuit who participated in the Aulavik Oral History Project and to Inuvialuit research assistants Agnes White and Shirley Elias. Translations and transcriptions of excerpts cited in this paper were done by Barbra Allen, Beverly Amos, Helen Kitekudlak and Agnes White. Special thanks to Nic Larter for sending me recent articles and reports that he wrote with John Nagy on muskox and caribou from Banks Island. The comments and suggestions of David Anderson on earlier versions of this paper were very much appreciated, as well as editorial help from Mark Nuttall. A shorter version of this paper was presented at the Twelfth Inuit Studies Conference at the University of Aberdeen, Scotland, in August 2000.

Notes

1   All quotes are from Nagy (1999). For each quote from the interviews recorded in 1996, the reference to the original English translation and transcription (Nagy, ed. 1999a, 1999b) is given. The first two capital letters correspond to the initials of the person interviewed. The same is done for translations of archival material (Nagy, ed. 1999c) and for written documents from archival sources.

7

Political Ecology in
Swedish Saamiland

Hugh Beach

In northern Sweden today (and indeed over the centuries albeit, in changing ways) two animal species, wolf and reindeer, in their relations to humankind and to each other hold key positions in a number of ongoing local debates. Both wolf and reindeer are major economic determinants of the livelihood of indigenous Saami pastoralists. Both are also powerful symbols, moving debates of resource conflict and compensation dramatically into larger discourses of principle concerning the relations between a native minority and the nation-state and between environmentalists and the Saami herders who have increasingly been cast in terms of 'eco-criminals'. Both Bjørklund and Wishart note similar processes of the criminalisation of indigenous livelihood traditions, in this volume.

> Despite the fact that Sweden has ratified conventions and is subject to the EU's habitat directive, we cannot provide our predators threatened with extinction the protection they need and their proper place in their native lands because of aggressive minority interests. The existence of predators is threatened mainly by trophy hunters (a perversion among the hunting group), illegal hunting and the extensive animal handling practices of the reindeer business. (Lindberg 2001: 21)

> To appease these Saami, the taxpayers are forced to give them compensation for lost reindeer. Should we not acquiesce to this, they would start exterminating our protection-worthy species.... My question to you Saami now is: how can you claim to be a nature-loving people today? (Löfgren 1999: 2)

Taken together, these two quotes point to the debates as perceived by the majority population, which generate a haunting political question: what obligations does the state hold toward the protection of the livelihood and culture of a small, indigenous people when balanced against the obligation to preserve Nature for all of humankind? While not presuming to answer this question, I will in the following pages seek to uncover spurious arguments in these debates and to refine their terms.

Just as, on the one hand, unrecognised consequences of Swedish herding policy are wrongly attributed to anachronistic Saami cultural values (Beach 1981), so, on the other hand, are unrecognised anachronistic cultural values of the majority population often cloaked in terms of hard science.

The reindeer is a key cultural symbol for anyone claiming Saami identity. Although, in Sweden, reindeer herding is a significant if not the major source of livelihood for only about 900 families, 2,500 people, or 10–15 percent of the Saami population, it is through this occupation alone that all the concrete rights to land and water accruing to the Saami people, either as an indigenous population or as individuals with immemorial rights, are allowed expression in Swedish legislation. Even Saami hunting and fishing rights, which are certainly as immemorial as that of reindeer herding, have been subsumed by herding legislation, so that they adhere as appendices to the Reindeer Herding Act and are available only to those actively engaged in the reindeer herding occupation.[1] Saami land rights based on the criteria of immemorial right rather than form of occupational livelihood were addressed in the Swedish Courts during the fifteen-year long Taxed Mountain Case (Cramér 1968–2002; Svensson 1997). Although the Saami did win some points of principle about the ability of pastoralists to own land, which might cause future claims to be successful in other areas, this case did not lead to any concrete changes in the contemporary legal framework nor in the actual Saami access to and utilisation of specific lands.

Moreover, and adding to the importance of the reindeer to the Saami, the right to herd reindeer (not necessarily to own them) is prescribed by Swedish law to fall only to those of Saami heritage, and all reindeer, regardless of owner, must have a registered (Saami) herder. Hence reindeer herding is a distinct Saami ethnic marker in Sweden, an ongoing traditional livelihood entailing the apprenticeship of skills and dwelling in Saami taskscapes, establishing deep-rooted continuity with past generations (Beach 2000a; Ingold 2000), and finally, it alone carries the formal legal expression of all practically recognised Saami land claims.[2]

Yet forces of change threaten the continuity of Saami herding traditions and redefine the content of Saami symbolic capital. A herd that is something for a Saami herder to be proud of is not simply one which brings high profits. It has beautiful animals in various proportions of age and sex classes (cf. Paine, 1994). More than just that, it also reflects the patterns of control exercised by its herders. Each animal and each herd entity as a whole, through its behaviour becomes the emblem of its herders. Despite such traditional Saami herd values, however, the pressures towards rationalisation and extensivity are inexorable, and herders have of necessity been quick to comply. Compliance is definitely not a painless adaptation.

Crudely drawn, problems with reindeer as perceived by the majority occur when this should-be tamed and controlled livestock becomes too wild. In contrast, the wolf becomes problematic both to the local population and to the ideological stance of their environmentalist protectors when it becomes too tame and subject to control. The different contributions to this volume clearly indicate the need to be culturally sensitive to different conceptions of such terms as 'wild' and 'tame'. Differences do not at all necessarily follow a Native/non-Native dichotomy. While so opposed in political position regarding natural resources, contemporary Saami and majority 'Swedes', it can be argued, share the same logic of what is 'wild' as opposed to what is 'tame'. Of course, what passing tourists might consider to be wilderness will be home ground for Saami herders dwelling in the land, but for both groups (at least as gleaned from the rhetoric of their spokespersons) the idea of 'wild' is that which is closest to a pristine nature devoid of human influence. It is just that the Swedes often do not *perceive* the Saami influence. However, to the Gwich'in (described in this volume by Wishart), 'wild' animals are those that have had their normal relationship with humans disrupted. Being 'wild' takes on the flavour of becoming temporarily insane, something quite unnatural.

A commonly held belief in Sweden today is that irresponsible herders have allowed such herd increase that their reindeer have exceeded rational reindeer quotas and threaten to trample and overgraze sensitive mountain areas into a rocky desert (Beach 1997; Ihse 1995: 14). The complaint becomes ethnically politicised immediately by media presentations announcing that the Saami herders are destroying the *Swedish* mountains (*Norrbottens Kuriren*, 9 December 1994, my italics). Claiming reindeer damage to newly planted saplings and general inconvenience and obstruction in the logging process by reindeer, numerous private forestland owners have recently brought Saami herding groups to court, in a series of notable cases, to contest the Saami right of traditional usage to graze reindeer on their lands (Cf. Sveg Case lower court verdict 1996). The accepted wisdom to account for such dire conditions is that the herders who own deer privately but who graze them on common lands are structurally trapped in the so-called 'problem of the commons', resulting in excessive herd growth. This, when compounded by modern 'high-tech' herding technology – snowmobiles, trucks, helicopters, motorbikes – results in ever-increasing herding 'extensivity' whereby the reindeer are subjected to short intermezzos of strong dominance but on the whole to diminished control. Large numbers with higher density of deer naturally imposes its own extensive herding pressure, for the animals are more prone to disperse in search of adequate grazing, and should grazing resources become ravaged, the imperative of commons dilemma moves with circular reinforcement, from the motivation of individuals to accrue even more surplus reindeer wealth, to that of securing the bare minimum number to survive within their livelihood. Even those of pro-Saami disposition, while they might absolve the herders of blame for the consequences of their actions (or inactions) because these are seen to stem from the logical and incontrovertible workings of the commons-problem structural trap, view the scenario as true nonetheless. Grudgingly they come to agree that curbs are demanded for a herding livelihood and further constraints for indigenous self-determination if Nature is to be preserved.

They generally fail to recall that the tragic element of the commons tragedy has been triggered far more by the massive exploitation of grazing lands by other competing land-use enterprises than by destruction of grazing by reindeer, and most importantly that the structural trap they correctly regard as a real problem need not be so were Saami demands for a 'reindeer account' policy permitted. As I have elaborated elsewhere, this is a system whereby reindeer could be removed from the grazing lands (i.e. slaughtered) and stored instead as wealth in the bank without incurring disadvantageous taxation regulations. Simply put, the reindeer could be slaughtered according to ecological factors but utilised according to a herder's need (Beach 1993: 112). In fact, when faced with herd sizes considered excessive, the punitive measures legislated in 1993 as addendums to the Reindeer Act of 1971 for oversized herds might better be exchanged for regulations devised so that slaughtering for bank storage became decidedly advantageous to the herders.

Similarly, with respect to the increasing mechanization of herding labour, it is rarely understood that this is today a trend driven by the stick far more than induced by the carrot. Market pricing, especially as reindeer meat has come to compete with the imported meat[3] of other deer species, leads to the condition whereby the herder needs an ever-increasing herd size simply to maintain his current living standard (a trend well established prior to the advent of costly high-tech herding, even if this has surely added impetus). Having reached a point whereby his household economy cannot be sustained on herding alone, the herder must invest his labour elsewhere for individual benefit, thereby relying more upon the collective work efforts of his Sameby colleagues to work in his herding interests as well as their own. Hence we find established a kind of tragedy of the labour commons (Beach 2000b: 198ff.), whereby the labour one removes from the collective herding effort causes less immediate economic harm to the individual in relation to the benefit of the same labour time invested in a private undertaking. Increased use of high-tech equipment compensates decreases in the herding labour force and labour time investment. The dynamics of such a situation leads directly to what I have previously termed 'the extensive spiral' in herding (Beach 1981). With the extensive herding spiral, the less time and labour one invests in herding, the wilder the deer become, the greater the reindeer losses, the less profits those remaining produce, and the less time one can devote to the livelihood, and so on.

These two trends, regardless of their ultimate origins or responsibility, integrate with each other. 'Rational herding', – that which by definition is to provide the greatest possible profit from herding within the bounds of environmental sustainability – demands a herding practice driven at full throttle with herds at the maximally permitted herd limits. At any one point in time variables such as rational herd age/sex composition can swing considerably according to market whims, such as those of meat classification with variable pricing and the meat-to-bone proportions favoured by restaurants. Yet, despite such ever-present shifts in reindeer husbandry variables to strive towards the elusive rational ideal, a precondition for attaining greatest rationality in herding is the sustainably *maximal* utilisation of the grazing lands. Herds established at this level, poised at the brink of excess, taken together with the trend towards increasing herding extensivity and mechanisation, rolling into the extensive spiral with diminishing herd control, are bound to breach established rational herd limits, unleashing cries of environmental degradation. In fact, one might claim that if this scenario does not repeat itself frequently, if herd sizes do not rise to challenge sustainable grazing limits, then herding is not ideally rational and the Sameby in question must suffer some negative herding pressure such as heavy predation.

The matter of herder enskilment is also vital to this equation (Beach 2000b: 204). Herders raised and active in a long-term period of extensive herding come to lack the skills of their predecessors. Admittedly they have new skills, but reversion to intensive-herding practices either of the past or of a modern variety is more than a matter of need and desire; it is also a matter of know-how. For the reindeer too, habituation to a herding system of greater intensive control is not something that can be reestablished in a season, a year or even a reindeer generation.

As we have seen above, harsh pressures towards extensivity and rationalisation characterise modern reindeer herding in Sweden today. At the same time, however, there exist a number of serious pressures calling for greater intensive herding measures. Public opinion demands animal handling practices involving little stress and closer human-reindeer contact. Many environmentalists regard low-tech but labour-intensive herding methods to be the least harmful for the land. Most importantly, should large numbers of reindeer become lost to predators following new legislation protecting species like wolves, herders must guard their reindeer more intensively.

The wolf is by far the predator species that carries the greatest symbolic capital for the environmentalist lobby. The wolf has returned from the brink of extinction in Sweden. In the early 1970s Sweden was the home of one purported lone wolf. Today the count is about eighty, most of them descendant from a single wolf pair. While Swedish researchers advocate a minimal wolf population of 500 to avoid the growing defects of inbreeding, the government follows a step-by-step programme which accepts the initial (more socially acceptable) goal of 200 wolves (Regeringens proposition 2000/01: 57). Towards this end, the wolf enjoys various forms of protection from hunting, which undergo continual negotiation.

According to the general Hunting Ordinance of 1987 (Jaktförordningen 1987: 905) a wolf might be hunted only if there is no other reasonable solution and if the maintenance of a viable wolf population in its natural area of distribution is not impaired. Even so, proper consideration must be taken for the size and composition of the wolf population. Such a hunt must be selective and controlled. The possible hunting of predators which might occur under the rubric of 'protective hunting', designed to protect property (livestock) from serious damage, cannot be invoked to hunt wolves, since the wolf population is considered too low and does not yet demonstrate a high enough birth rate to tolerate decrease. Security for wolves within the borders of national parks is even greater. According to the Hunting Law:

> Has a bear, wolf, wolverine or lynx attacked and injured or killed tame livestock and there is reasonable cause to fear a new attack, the owner or caretaker of the livestock may without hindrance … kill the attacking animal, if it occurs in direct connection to the attack. Such hunting is however not permitted within a national park. (Jaktlag SFS 1987, 28§)

With respect to the wolf, the general law of self-defence for oneself and one's private property has been constrained. A herder can no longer, without serious reflection, inflict injury on a wolf that is attacking his reindeer. A herder can try to round up and relocate his herd, but he cannot retaliate against a wolf aggressor upon its first attack. Retaliation can occur only when the attack is repeated and even then cannot involve chasing or tracking with any kind of motorised vehicle. Other legal paragraphs prohibit anyone to travel by motor vehicle together with a fully assembled weapon, or for the weapon to be used in the vicinity of the vehicle. In short, the herder must in all probability be present during the first attack, when he must refrain from offensive measures, and certainly during an ensuing, second attack, when his response must be on foot, immediate and face-to-face. These are hardly the characteristics of wolf-hunting practices, which rarely occur directly under the nose of a herder.

The biodiversity argument invoked in favour of maintaining a wolf population takes as its point of departure that each species has evolved to play a role in the total holistic ecological system, and to eliminate one player impairs the survival of the others and handicaps or threatens the survivability of the whole. Yet for predators at the top of the food chain, this argument has vulnerable points. They themselves hardly service other species significantly by their own meat as food; instead, their beneficial ecological role for the larger whole is through their killing services, combined with the 'trickle-down' distribution of the carcasses of their prey by which wolves come to circulate resources back into the system, and through their services as selective breeders for other species, culling out inferior stock.

One must question if a wolf population of even 500 individuals would contribute in any meaningful way to these ends. Moreover, their selective breeding benefits, certainly to the reindeer population at least, are obviated by the slaughtering practices and selective breeding practices of the herders themselves. Instead, the wolves can come to destroy the herders' selective breeding efforts by killing valuable reproductive stock. Inferior reindeer have already been culled out with good husbandry methods. While the biodiversity argument might in principle retain some limited validity for wolves in the reindeer-herding regions, the presence or absence of wolves even in the maximal numbers considered does not carry great practical ecological significance of a positive kind. Yet, there are still other arguments for preserving a viable wolf population despite the enormous economic, social and political costs involved.

Aside from arguments based on the role any species plays in the larger ecological system, the environmentalists advocate wolf preservation on the grounds that Sweden must fulfil its international obligation to protect endangered species per se. When the Saami point out that internationally the wolf is far from endangered, with many thousand nearby in Russia, the point is made that Sweden must do its part to maintain the particular Scandinavian wolf type. And yet, when the pro-wolf lobby discusses the goal of 500 wolves in Sweden against the backdrop of the inbreeding threat, it is argued that this potential problem might be overcome as fresh wolf genes are added to the Swedish pool by wolves crossing from Russia into Sweden via Finland:

> For genetic exchange to occur between the Scandinavian and the Finnish–Russian wolf populations, wolves must be able to move from the border with Finland to the area where the main part of the Scandinavian wolf population is located. Such an exchange is of highest importance to ensure that the Scandinavian wolf population does not become genetically impoverished. (Regeringens Proposition, 2000/01:57, p. 42)

In January of 2002 a rare drama unfolded: a lone wolf had wandered across the Finnish border into northern Sweden. Once discovered by the Swedish Environmental Protection Agency (SEPA), the animal was tagged with a radio collar two days later and for many months has been monitored in its every move. The wolf was put to sleep, and DNA samples were taken which confirmed that this wolf was genetically close to wolves found in eastern Finland and Russia and is, therefore, according to the SEPA, 'a highly interesting individual' (2002). Helicopters have followed it by air, and snowmobile-mounted protectors have kept a watchful eye to ensure that it will not be shot. The Agency points out that the radio collar should help the Saami herders to move their herds out of harm's way. Already the herders estimate the cost in extra work occasioned by this wolf to be approximately US$3,000. In a decision taken on 11 March 2002, the Agency denied petitions to shoot the wolf or to move it out of the herding area with the declaration that '...the goal for the distribution of the wolf population shall be that it spread itself over the land *in a natural way*' (my emphasis).[4]

When the ideal of gaining freedom from the impending inbreeding problem is considered in relation to: (1) the time-scale of the government's step-by-step programme and the huge public 'NIMBY' ('not in my backyard') outcry against even the existing wolf population by local people, not to mention the goal of 200 or the dream of 500; and (2) the infrequency of genetically attractive Russian/Finnish wolves straying permanently into Sweden, the idea has been broached of flying in wolf sperm from Russia. Hence, a genetically healthy wolf mix could be obtained without prejudicing a hot debate with added wolf bodies. Interestingly, and as I have argued elsewhere (Beach, 2001: 97), implementation of such a concept would subject what can justifiably be called 'wolf-managers' to the same critique as that held against Saami reindeer managers, viz. that theirs has become an 'unnatural' high-tech-dependent enterprise of little economic benefit to the majority, a kind of hobby maintained by the Swedish taxpayer. In fact, the remarkable helicopter-and-radio-monitored surveillance operation around this most recent wolf immigrant displaces this point as one of mere conjecture. And, ironically, the culling of eastern wolf sperm for implantation in Sweden, which was at first launched as a humorously provocative extreme argument to pin down a point of principle, is no longer met with a dismissive smile; the SEPA, now seriously worried over the safety of their latest immigrant wolf, the improbability that it will travel far south to mix with the permanent Swedish wolf stock, and the extreme expense of the surveillance venture, are now contemplating culling his sperm.

It seems that endangered status does not hold in itself as an argument to preserve a viable and free wolf population in Sweden, and the argument based on Scandinavian typing defeats itself by its own anti-inbreeding survival plan of 'foreign' wolf import. However, there are still other arguments involving ethical and moral principles, which hinge largely on what is considered to be true and pristine Swedish wilderness. A deep-seated and understandable conviction of the environmentalist ethic is that at least somewhere within the nation's borders, humankind must command a halt to its destruction of Nature and leave the land unbent by human purposiveness. Of course dilemma ensues, for this too is a human purpose, and just as in the case of discriminatory affirmative-action employment programmes to right the wrongs of previous discrimination, environmentalists seek by means of the oxymoron 'wilderness management' to regain Eden lost.

But what exactly is the Nature targeted as pristine? Does it include modern reindeer herding? Or, if there is to be herding at all, should it be only of the low-tech variety? Did Swedish Eden have eighty, 200 or 500 wolves? Nature becomes consciously negotiated in direct proportion to the conscious command we have over it. A capability unexercised becomes as decisive a human (in)action as its implementation. In effect, doing nothing of what one *could* do becomes an instance of wilderness management, whereas previous to such ability of control and beyond its reach, all that occurred was 'natural'. Hence Nature itself shifts along an intensive–extensive continuum of control, analogous to that of herding, depending on our knowledge and abilities, regardless of its actual environmental state. This point is far from being simply philosophical. A concrete example from Saami herding reality and minority politics will serve to illustrate its bearing.

Few things can be more illustrative of the conception of Nature than the discourse around compensation funds or subsidies, when distributed to ease the suffering of what is considered to be natural disasters or unavoidable misfortune. Funds paid for the reindeer that herders lose to predators are commonly perceived of and counted as a subsidy – to the reindeer herders – by the state. One recent investigation concluded that the reindeer livelihood was subsidised by fully 178 percent of its production (Johansson and Lundgren, 1998). Naturally this enrages the common Swedish taxpayer who thereby seems to carry the burden of sustaining this troublesome little minority. Adding insult to injury, Saami herders, it is argued, insist on engaging in a counterproductive business, a kind of cultural hobby that impedes industrial development and full employment in the north; and, finally, to cap it all, Saami livelihoods are supposedly destroying the environment.

Certainly the funds are a partial compensation to the reindeer managers for their losses, but they are simultaneously a subsidy not to them but to what we might justifiably dub today the 'wolf managers', to establish conceptual parity with reindeer managers. Should the government buy fodder to feed cattle threatened by starvation, this would be considered to be a subsidy to the cattle owners. However, when the government buys predator food (reindeer) to feed predators, this is considered a subsidy to the reindeer herders. If the Swedish accounting system were made to reflect expenditures whose true goal is to sustain the lives of wolves and other natural predators, then the amount posted at the door of the Saami herders themselves would shrink correspondingly. This would undoubtedly be fairer and more representative of the Swedish government's actual commitment to the Saami herding culture as such.

The reason compensation money is counted as a subsidy for the reindeer herders lies, I believe, to a great extent in what we consider natural, and the priority of Nature over human claims. We think it only natural that wolves kill reindeer, that wolves should abound in reindeer pasture lands, and that the state is doing a good deed by helping the herders absorb their losses. But it is also natural for the herders to kill wolves to protect their property; and this the law does not permit. Historically, the dramatic decline of the wolf population in Sweden over the centuries has not been caused primarily by Saami hunters, but rather by the spread of Swedish settlement. The establishment of a

viable wolf population in Sweden today is a feat of human environmentalist engagement and government legislation.

Finally, it might be appropriate to quote here the ultimate expression of globalised political conservationist argumentation for wolf pro-
tectionism in Sweden. When asked at a conference about predators
why Sweden should protect endangered predators, the Minister of the
Environment, Kjell Larsson, replied:

> It is not a burden that we have accepted to do this. It is a small, small
> price that we have to pay, but it gives us the possibility to participate in
> international discussions about the preservation of endangered species.
> We can put pressure on other countries if we ourselves take responsibil-
> ity for our threatened species. (Larsson 2001)

Cynical herders are among the first to confirm that it is indeed only a
small, small price that the government is willing to pay them for all the
damage caused to their livestock by predators. Recently, however, compensation sums have improved, even to the point of incorporating
what is commonly considered to be a kind of 'bribe factor' to induce
the presumed future guilty Saami herders to refrain from killing endangered predator species illegally (Dahlström 2003: 483).

The policy for predator compensation has been totally revamped. A
new system of compensation agreed upon and proposed by both the
herders and the Swedish Environmental Protection Agency is that the
herders receive payment not for the number of animals lost, but
instead according to the number and kind of predators with their dens
on Sameby territory. Of course, predators do not necessarily care
whether they are hunting on the territory of the Sameby receiving compensation for them, or if they leave their home area to hunt in the
neighbouring Sameby. In such cases, 'shared' predators elicit coeffi-
cients for multi-Sameby compensation. However, the government did
not finalise in legislation the new system it called for, once appraisals
of reindeer losses indicated just how much livestock the predators
destroyed, i.e. how much it would cost to convert the wolf from being
an economic burden on the Saami into being instead a source of economic gain. At first the government appropriated only a part of the
funds known to be necessary to compensate herders for their losses.
Nonetheless, the new form of compensation, based on living predators
rather than dead reindeer, brings Sweden closer to the principles advocated by eco-tourism.[5] The presence of predators introduces the possibility of their becoming economically attractive to the herders
(especially if intensive herding methods were to minimise reindeer
losses). Such a form of compensation, it is believed (provided it is ade-

quately funded), would convert into predator protectors those who might otherwise want to disregard the law and to snipe at wolves.

It is assumed at least that the new higher rates paid for predators will stop herders from killing them illegally. That amount of compensation, which exceeds real losses constitutes the so-called 'bribe sum'. The reindeer herders' political organisation, SSR, has officially objected to any such bribe sums, dubbed as 'nature value' grants placed on the predators, since it has been so obviously connected to the unwarranted conviction that trigger-happy herders will hunt illegally without compunction. Hence, if predators are spared, herders are viewed as having taken the 'bribe'. If not, it will confirm the idea that herders are irrationally rabid predator haters for whom even superadequate compensation is meaningless. Either scenario confirms the assumed guilt of the herders, their eco-criminality.

As human control and even possible control over the environment increases, we are forced to reevaluate what we conceive of as Nature, and, as a consequence of this, how we conceive of and relate to indigenous livelihoods. Ironically, the same kinds of international covenants and declarations devised by removed, often foreign, politicians, dedicated to altruistic principles of conserving the unique and threatened, and invoked to 'save the wolf', may come to be invoked to 'save the Saami' as well. Already it is difficult to conceive of an indigenous population anywhere whose traditional livelihood is as carefully controlled and managed as that of the Saami in Sweden. New and simplistic ecological arguments for further increases in regulation (whether born of good intentions or ulterior motives) further constrain Saami self-determination nonetheless, spawning among Saami terms such as 'eco-colonialism' or simply 'ecolonialism'. To the Saami, Swedish environmental concerns are unavoidably tainted by hypocrisy. It is the majority Swedish society which has permitted the massive exploitation of northern natural resources, forced herding into a tight corner, and which now castigates and fines small-scale Saami livelihoods for being ecologically nonsustainable and threatening to that terribly diminished 'wilderness' which the majority wants to maintain for its own needs of tourism and nature romanticism.

Ecology in practice is unavoidably political and never purely scientific. Goals of 'sustainable development' beg the questions: what is to be sustained, and for whom? There are an infinite number of long-term sustainable ecosystems that can be promoted in a given region, which is a political question. What can be termed 'vulgar ecology' tends to cloak the role of human purpose in conceptions of Nature. It is a perspective readily revealed by the reductionistic, monetary metaphors it employs; one should live on the 'interest' and not deplete the 'capital' of natural resources. Supposedly, if one follows this rule of thumb, Nature (or whatever ecosystem has been targeted by human purposiveness, for example 'wetlands') will be sustained. However, in the monetary metaphor, even if amounts of it change, money is a constant. One is sustaining it, increasing it or depleting it. But ecosystems do not work this way. In whatever way they are being utilised and to whatever degree, they thereby alter character (not just quantity).

Acknowledgement

The Swedish Research Council has financed research for this paper. I am indebted to the efforts and insights of the other members of the research project team, Åsa N. Dahlström and Carina Green.

Notes

1   Note, however, that in 1993 the Swedish state confiscated the Saami exclusive small-game hunting right on Crown lands, claiming that the Crown holds a parallel hunting right as owner of the land – although the state has been unable to prove ownership. Nonetheless, the Crown's parallel right has been implemented, and while the Saami small-game hunting right continues, it is no longer exclusive and cannot be orchestrated by the herders to be conducted in a way harmonious to the spatial and seasonal calendar of herding activity (Beach 1994).

2   In Sweden, it becomes necessary to distinguish between having a right and being permitted to practise that right. For example, in what appears to be a broad ethnically based right, *all* Saami have the right to herd reindeer according to the first paragraph of the Swedish Reindeer Herding Act. However, ever since 1928, only those Saami who are members of the Sameby herding collectives (of which there are about fifty in Sweden) have the right to *practise* this herding right on Sameby territory. These Samebys form 'closed shops', as the membership of each Sameby is controlled by its existing membership, who are not at all prone to take in newcomers to share further their limited grazing resource and tightly regulated total 'rational' Sameby herd sizes. For an account of Saami herding conditions and legal constraints in Sweden, see Beach (1981, 1994 and 2000b). For a detailed historical account of the development of Swedish legislation concerning the Saami, see Cramér (1968–2002).

3   Interestingly enough, one of the effects of the Chernobyl nuclear disaster was to establish firm mule-deer meat imports from New Zealand to substitute for the loss of reindeer meat on the Swedish market due to fears of contamination.

4   Recent news alerts of this wolf indicate that it is moving farther north, to the chagrin of the SEPA, which hopes it will bring its genes south to be united with the rest of the Swedish wolf gene pool. Moreover, movement by this wolf in any direction out of the maximally protective zone afforded by the Laponia World Heritage Site's complex of national parks causes anxiety to those bent on his survival and breeding success. Yet

the herders who knew of its presence for months before the SEPA, and who might have surreptitiously disposed of this wolf, did not.

5 Saami political discourses about tourism, which I have encountered in Sweden concern the issue of who shall benefit and how it can be pursued so as not to hamper or disrupt reindeer management. Placing it under Saami control in herding rangelands would favour proper integration with herding practices and also provide Saami with employment opportunities. Practising Saami herders are likely to express priority for the wellbeing of the animals in their care, but such attitudes are distinctly different from the kind of nonintrusive respect of nonhuman species and the consideration for the landscaping habits mentioned by Wishart in this volume, as illustrated by the Gwich'in.

8 Saami Pastoral Society in Northern Norway: The National Integration of an Indigenous Management System

Ivar Bjørklund

Background

In 1992 the Norwegian government delivered a report to Parliament, in which it concluded that 'the law (regarding reindeer husbandry) has not worked according to its intentions. [It] has not been able to secure balanced resource management and viable adaptation'.[1] These are rather harsh words for a governmental report and they certainly beg some hard questions. In this chapter I will therefore take a closer look at why this policy has gone wrong and ask what the consequences of this failure are. The answers shed some light on the viability of indigenous management systems in ecological terms and the consequences of the economic and political integration of such systems into the national state.

Saami reindeer herding has seen significant changes through the centuries, but some ecological basics remain constant. In management terms, these are reflected through concepts such as mobility and flexibility. However, a wide concept of variation is also important. Any overview of reindeer herding in Norway will tell us, for instance, that twice a year over 150,000 reindeer move between winter and summer areas in the northern county of Finnmark. From this it appears obvious that reindeer herding has a lot to do with variation in terms of pasture. We must also remember that variation in terms of animals also plays an important role. As Figure 8.1 tells us, there are several categories of animals in a herd. Different kinds of animals need different kinds of management, as the herders move the animals according to varying grazing conditions. The pastoral task is to obtain the optimal relation in time and space between pasture and animal (Bjørklund 1990). This has, of course, been the everlasting problem – and solution – in pastoral adaptations in northern Norway.

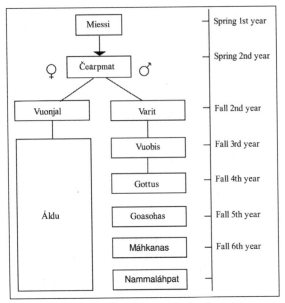

Figure 8.1 Traditional reindeer categories by sex and age in Saami.

In Saami pastoral society every animal is owned individually. The animal belongs to an adult, or even a young child, who cuts his or her mark in the ear of the animal. These earmarks are significant cultural devices that tell stories about social relations among the owners (Bjørklund and Eidheim 1999). One consequence of such individual ownership is that one must always move the herd in such a way that one takes care of the interests of both animals and owners. This constant attempt to mediate the relationship between pasture, animals and their owners must, by necessity, be organised in specific ways. It is this organisation we could call an indigenous system of resource management. In the following I will give a short outline of how this system has been based upon pastoral knowledge and organised through a Saami cultural institution called the *siida*.

Cooperation between reindeer owners is organised through kin relations (Blehr 1964), with relations among the sibling group being the most important (Pehrsson 1957; Paine 1970). The term *siida* refers to a group of reindeer owners who live and migrate together, and to the herd of reindeer owned and herded by them. Because of the varying grazing conditions through the year, the demand for herding tasks and labour will also vary. Consequently, the *siida* changes size and composition through the year, as the pastoralists divide and regroup their herds. In this way, the *siida* represents a flexible cooperative unit between people and animals. By dividing and combining herds and personnel throughout the year, Saami reindeer herders attempt to achieve the optimum relation between animals, pasture and labour (Figure 8.2).

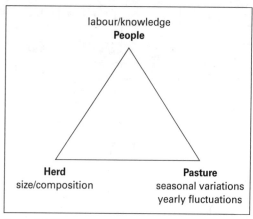

Figure 8.2 Diagram representing how the *siida* achieves a balance between people, animals and land.

For instance, the available pastures in a given calving area might not be enough for the number of animals who were together in the wintertime. It follows that a reasonable strategy is to split the winter herd into smaller herds, each moving into different areas and seasonal pastures. As the herds differ in size throughout the year according to the various grazing conditions, so also does the demand for herding tasks, knowledge and labour. The strategy of the pastoralists is never to be in a position where the size and composition of the herd is not in proportion to the available labour and pasture. If such a situation occurs, individual herders will try to withdraw their animals from the common herd and join other herding units – according to kinship relations and available pasture.

This management system no longer operates today as it did a generation ago (Pehrsson 1957; Paine 1994). Instead, since the late 1980s, this Saami resource regime has been integrated gradually into the Norwegian national state. These integrational efforts have taken place along three dimensions:

1 The ideology of the welfare state, which prescripts a levelling of income and economic welfare for all. The state is supposed to be the caretaker of the interests of any member of society and the basic political goal is to provide these members with a fairly equal amount of social and economic welfare. This is done through a rather complex system of laws, regulations and political negotiations. As for the pastoral Saami, before the 1980s they were more or less outside the corporate channels of the state and this situation constituted a problem for the national authorities. In addition, statistics were presented as proving that the income of the pastoral Saami in monetary terms was way below the national average. Furthermore, many politicians referred to the pastoral society as rather hierarcical and unjust, because the wealth in terms of animals was unevenly distributed among the owners.

2 Conflicts regarding the use of land came apparent throughout the 1970s. A growing number of land-use conflicts appeared in the reindeer herding areas because of the building of new roads, hydroelectric dams and military installations. This development led to strong protests from the reindeer pastoralists and some of the cases were taken to court. This development culminated in the dramatic case of the Alta-Guovdageaidnu hydroelectric project around 1980 (Bjørklund and Brantenberg 1981, Paine 1982)).

3 At the same time, new technological innovations were embraced by the pastoralists. Snowmobiles and, later on, motorbikes and four-wheel-drive vehicles made new herding techniques possible, but also generated a growing need for money. Governmental housing programmes and a fast-growing supply of consumer goods only contributed to an expanding cash economy and its stresses.

All these processes led to a situation in Norway where it was considered politically important 'to do something' about reindeer herding. In governmental language this meant turning it into a national economic sector with specific aims and rules regarding concepts like 'modernisation' and 'rationalisation'. Because of the growing number of animals at the end of the 1970s, and the perceived impacts of reindeer on the environment, governmental experts looked upon reindeer herding as a living proof of the tragedy of the commons and argued for governmental interference. Others – among them quite a few Saami – thought of it as a source of income which could be made considerably more profitable through governmental intervention and control and, not the least, distribution of subsidies.

This development led to a special economic agreement in 1976 affecting all reindeer herders, followed by a new law on reindeer herding two years later. The main intention behind the law and the regulations specified in the agreement was, in national economic terms, to transform the pastoralists into meat producers and thereby generate economic growth in the sector of reindeer herding. This was to be done by (1) reducing the number of animals, and (2) regulating the herding activities. The rationale of the agri-economists was that fewer animals inevitably led to bigger animals and, with the help of more efficient forms of herding, the production of meat would expand while the "carrying capacity" of pastures would not be exceeded (Government of Norway 1985). The practical consequences of this policy were the introduction of an upper limit regarding the number of animals allowed in each district and an extensive set of regulations to rationalise and modernise the herding.

From then on reindeer pastoralism has been a management system in transition, not unike the situation described by Patty Gray (this volume). This transition has to do with the fact that Norwegian political institutions have now taken control of the pastoral factors of production – a control that earlier, as I have already suggested, was exercised through Saami cultural institutions. This situation is not unique to Norway. As Beach (1997) shows for northern Sweden, Saami reindeer herding is under pressure from legislation to make it conform to the Swedish state's view of what constitutes profitable business. However, as reindeer herding management practices become 'business-like', Swedish Saami also fall under criticism by environmentalists for abandoning what they regard as a traditional lifestyle. This process of integration in northern Scandinavia is generating significant political and social turbulence. In the following I shall examine more closely the steps and substance of this integration as it takes place in Norway, and consider whether something is to be learnt from what is happening to and within this indigenous resource management system.

The Herd

My concern here will be how governmental interference affects the size and composition of the herd. In order to obtain government subsidies, each reindeer herder must slaughter and sell a certain percentage of his or her herd. In addition, for many years there was an extra bonus for those who would slaughter calves – which was an idea strange to pastoral values. The government thus interferes directly with husbandry decisions (Paine 1964), and attempts to regulate which animals to slaughter and which to let live. In local communities, the ability to make this kind of decision was always considered proof of being an independently minded reindeer owner. In a sense, being in control of the life-cycle of the animal is what pastoralism is all about. The aim of this policy was to reduce the grazing pressure in the winter areas. This policy, however, immideately led the pastoralists to change the composition of their herds. They found it profitable to reduce the number of bulls and increase the number of females, thus being able to produce more calves than before. Such changes generated a growing number of reindeer altogether – which increased the detoriation of pastures and the general problems of management.

The fact that the size and composition of the herd are now regulated by government economists has had serious consequences for the maintenance of pastoral knowledge. As one can see from Figure 8.3, the number of animal categories has today been drastically reduced. The system of subsidies has made it profitable to have only four or five categories in the herd. Also, mechanisation has eliminated the need for draught animals. According to the government's scheme, the ideal winter herd today consists of almost no calves and very few bulls – not withstanding the problems that this introduces for reproduction.

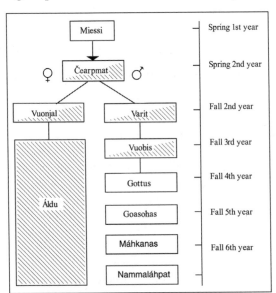

Figure 8.3 Diagram representing how the number of reindeer categories has been recently reduced in northern Norway as a result of subsidies.

This situation, of course, makes the herders much less flexible when it comes to manipulating the herd. For instance, the greater the proportion of bulls in a herd, the easier it is for the herder to separate the herd into two parts according to gender, and to manage the herds in different ways and directions if, for instance, pasture conditions should make this necessary. Thus, the current herd composition makes flexible herding strategies difficult both in regard to herding and to husbandry. For instance, winter grazing has become less efficient due to the lack of bulls penetrating frozen snow and thus giving grazing access for weaker animals. Husbandry decisions are also much more fixed in the current situation, because the reindeer owners have fewer options when it comes to the slaughter and selling of animals.

We can also tell from Figure 8.3 that the 'modernisation' of the herds inevitably has a cultural dimension, too. It has greatly reduced the traditional knowledge – and thus the vocabulary– used by the herders when they speak about reindeer. There exists a vast amount of ecological and social knowledge regarding the different categories and kinds of animals and their behaviour. In the Saami language, this knowledge is reflected in different ways to classify reindeer according to gender, age, colour, shape, horns, behaviour etc. Today there is a growing interest internationally in the importance played by this kind of 'traditional ecological knowledge', as it is often called in academic presentations.

Pasture

A rapid growth in the number of animals has been one of the most obvious consequences of the integration of Saami reindeer pastoralism. For statistical reasons it is very difficult – not to say impossible – to estimate the number of reindeer in Finnmark through time. Reasonably correct numbers are only available for the last ten years or so. However, it has now been documented how the relatively strong growth over the last fifteen years is a direct consequence of subsidy policy (Sara 1990). Subsidies have reduced the need for slaughter, because the pastoralists are reacting according to the kind of rationality described by Chayanov more than seventy years ago: production decreases as cash income increases.

The political debate in Norway today focuses solely upon the number of reindeer. But a growing number of animals also generates a growing number of herds because of management necessities and cultural practices. It is in the number of herds that we find the explanation behind the pastoral turbulence of today. Because herds, as mentioned earlier, are separated according to seasonal conditions and labour demand, more herds have direct practical, social and ecological consequences.

Firstly, the growth increased the possibility for random mixing of animals, which then had to be separated again. The increasing amount of herd separations did generate social conflicts, since such operations inevitably involved questions regarding ownership and responsibility. And furthermore, the ever-repeated separations had serious effects on pasture conditions. A separation today involves a significant amount of motorised activity – the animals are herded together by the use of motorbikes and/or snowmobiles and rushed into large corrals. When this takes place at a time of the year when the ground is not covered by snow, it results in the destruction of lichen pasture. Being dry in the summertime, lichen is extremely vulnerable to any kind of wear, be it from motorbikes or reindeer hooves.

To reduce the problem of mixing herds, long fences have been built all over the spring, summer and fall pastures. These fences have an impact upon Saami management practices because they not only separate herds, they also separate pastoral areas, making it a criminal offence to use pasture not assigned to herders through the legal system of the state. This situation is reinforced through official regulations stipulating when herds can enter and leave a given area. On top of this development come the ongoing encroachments on land. Tourism, roads, power lines, and so on, not only reduce available pasture, but have a tendency to close off areas which are of strategic importance to the pastoral herding strategy.

The growing numbers of animals and herds, together with the reduction of available pasture, have strongly reduced the most important asset of the pastoralists, namely flexibility. It is now becoming more and more difficult to cope with variations in climate or pasture conditions. Traditionally, the reindeer pastoralists were able to mediate the carrying capacity in a given area. It is extremely important to bear in mind that in this context the concept of carrying capacity is not a fixed size, as many biologists would argue. On the contrary, it is something which – within given biological limits – can be manipulated through management practices and herd composition. The carrying capacity of any area is something one evaluates and then mediates if necessary.

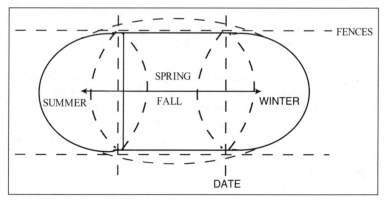

Figure 8.4 Diagram representing how the carrying capacity of any area is evaluated and mediated by Saami reindeer herders in northern Norway.

Figure 8.4 gives an idea of these dynamics. If, for instance, pasture becomes too scarce in the summer area because herds are expanding, herders might leave the area early and keep the animals longer on the autumn and winter pastures. Or, if conditions in the autumn become very difficult due to climatic fluctuations one year, one solution might be to move the herd through a neighbouring area, especially if it is temporarily vacant. This kind of flexibility has characterised Saami reindeer herding as long as it has existed, but is today becoming more and more problematic. Fences, pasture regulations and a growing number of herds and animals have strongly reduced the possibility of such strategies.

Personnel

Today, working units are facing problems regarding the management of both knowledge and labour. The law of 1978 introduced an official permit (*driftsenhet*) for the right to be a reindeer owner. Most Saami at the time saw this as rather offensive to their values and practices, in which capability, talent and kinship relations were determinating factors. The distribution of permits was also to a certain extent arbitrary and gender specific. Many owners – mostly female – were left out. And from the middle of the 1980s, no more new permits were given because the Department of Agriculture thought it necessary to reduce recruitment. From that time on herding permits could only be obtained by succession within the family.

According to the letter of the law, the recruitment of herding units was now under control of Norwegian political institutions. Traditionally, Saami cultural arrangements had taken care of recruitment into pastoral society. Animals were allocated to children during certain ritual occasions. When one received a reindeer for the first time, for example when a child was baptised or confirmed, one also received an earmark, making one a de facto reindeer owner. Along with the gift also came the responsibility of being a reindeer owner. Children learned how to take care of their animals and were thus socialised into the world of reindeer pastoralism. When the time came to marry, both spouses were in possession of knowledge and enough animals – together with the animals given to them as wedding gifts – to make it possible to establish themselves as their own husbandry and perhaps herding unit.

The fact that recruitment today depends upon legal rules of Norwegian origin, and thus political circumstances in Norwegian society, has profound consequences for the pastoral management of both knowledge and labour. As for knowledge, the traditional way of recruitment meant that without knowledge and skills one could not establish oneself as a reindeer owner. But within today's administrative system, there is not necessarily a connection between knowledge and recruitment. It is not skills and knowledge that decides who gets a permit, but other circumstances, such as political decisions and rules of inheritance within Norwegian society. It has always been a basic pastoral value that all children should be able to try out their interest as herders. But it was also very well accepted that not all had the abilities to succeed. Today, it is not only one's competence and skill, but rather an official permit which makes a person able to succeed in economic terms.

Concerning the consequences for labour, the *siida* is experiencing a loss of flexibility and facing new sources of conflict. The herders are now grouped into A- and B-teams, so to speak. Because of the subsidies that flow from the economic agreement, the herding permit has become a valuable asset in monetary terms. Only those with a permit get an annual cheque from the government; those without get nothing. Understandably this situation creates conflicts within the working units – the *siida* – because all members do more or less the same kind of work.

The loss of flexibility goes on both within and between the working units. If a person does not have a permit, but works as a herder and as his own owner- as quite a few people do – then this person is quite dependent upon somebody in the *siida* who has a permit. In legal terms, he[2] is the caretaker of your animals. This situation makes it very difficult to leave the *siida* and turn to someone else; legally one is stuck with the permit-holder whether one likes it or not. Thus the composition of the *siida* becomes more or less fixed regarding membership.

But also the relations between *siiddat* have become less flexible. According to the old administrative infrastructure, the summer pastures were formally divided into 'Herding Districts' (*Reinbeitedistrikter*). But the winter pastures are still organised and used according to Saami customary law and traditional rules. Within each 'Summer District' one might find up to six or seven *siiddat* which form different working units in the winter time. Now, these siiddat have to cooperate within the 'Summer District', but the borders between the districts do not always reflect common interests among its members. Some siiddat might have overlapping management strategies, which represents potential conflicts because the district has to act as one entity in all matters concerning common pastoral affairs.

For instance, membership in a 'Summer District' sets the number of reindeer which must be slaughtered in order to fulfil the regulations. The District has a collective responsibility to make sure that everybody slaughters the amount specified in the economic agreement with the state. If one person slaughters less, the District might not be able to fulfil its quota and subsequently no subsidies will be paid to the district as such. If one bears in mind how highly individual decisions are valued in this society – especially when it comes to reindeer husbandry – one can easily imagine the dilemmas coming out of this enforced collective responsibility.

Destructive Trends

A concluding remark to this essay could very well be the government statement cited in the beginning, namely that all efforts to integrate Saami reindeer herding into the Norwegian welfare state have ended in failure. The strategy of the herders has been to use any available means to remain in their pastoral adaptation. During these efforts, the character of their management has changed.

Because access to pasture has become increasingly less flexible and the opportunity for traditional cooperation is reduced, the control over individual animals has now become less critical. Much more important is the control over the herds as such. Thus the animals do not have to be so tame any more. The herders have therefore developed management forms where they only exercise control over individual animals when it is necessary. These occasions are when herders earmark the calves, separate the herds, and select animals for slaughter.

It goes without saying that this development also implies a loss of knowledge related to the single animals and their habitats. The critical knowledge these days concerns herd management and the use of modern technology, not behavioural or biological characteristics among individual animals. The ongoing reduction of animal categories, as presented in Figure 8.3, is a reflection of this development.

It is in this context that the extensive use of technical equipment must be understood. It is the necessity for keeping control of the herds that motivates many herders to invest half of their income in expensive technology such as snowmobiles, motorbikes and mobile nylon fences. It is this equipment – not to say its use – which today constitutes the Norwegian image of what Saami reindeer herding is all about.

In sum, the way Saami pastoralists adapt to the policy of integration is by accepting what could be used in their pastoral adaptation and rejecting the rest of the policy and its devices. It is these strategies that, over time, have created destructive traits, not only ecologically, but also in social terms. As demonstrated by the breakdown of the traditional working unit – the *siida* – this development has both an ecological and a social dimension. Many pastoralists will find themselves in a double-bind: a herder who wants to act according to customary knowledge and law will bear the risk of being punished one way or the other. He might either become a criminal, legally speaking, because laws and regulations often exclude established and well-proven forms of management, or he may be punished economically because the policy of subsidies only pays for those who manage their herds the way the state wants them to – and that is a way contrary to most Saami values and customs.

Notes

1 St. meld. nr 28 (1991–1992) *En bærekraftig reindrift*, s.67, Landbruksdepartementet, Oslo

2 In 1994, 87 percent of the registered reindeer owners in the county of Finnmark were men.

9 Chukotkan Reindeer Husbandry in the Twentieth Century: In the Image of the Soviet Economy

Patty A. Gray

Introduction: Post-Soviet Chukotka

The collapse of the Soviet Union in 1991 had far-reaching effects that precipitated social transformation throughout Russian society. Although most of the information that reaches the West concerns the more visible locales, such as Moscow, the most cataclysmic changes occurred in rural areas far from Moscow, about which little is generally heard. This paper concerns one particular region of the North that is located as far from Moscow as one can go in Russia – the far north-eastern Chukotka Autonomous Region. It concerns, moreover, one of the key economies of the North, reindeer herding, which is also one of the economies most severely affected by the post-Soviet transformation.

As everywhere in the Soviet North, prerevolutionary Chukotkan reindeer husbandry was a private (i.e., non-state) occupation of indigenous communities. It underwent collectivisation beginning in the 1920s and by the height of the Soviet period had been transformed into a branch of the Soviet economy. Today, by all accounts, Chukotkan reindeer husbandry is in a state of collapse, as deer head-counts plummet, dwindling herds are merged, and entire herding operations liquidate. Most agree that Chukotka represents the worst-case scenario in all of the Russian North for the fate of post-Soviet reindeer herding (see Krupnik 2000a: 53). Part of my goal in exploring these issues is to understand why this is so.

Assumptions about the Soviet system and its legacy often take one of two extremes. Stated in oversimplified terms, they are:

1 The Soviet system ('Communism') was bad, and democracy/privatisation will fix everything (for example, it can be the salvation of Chukotkan reindeer herding by providing an outlet to markets for reindeer products).

2 Privatisation was ill-conceived and badly executed, and is to be blamed for all that is bad now, especially in rural areas (for example, it precipitated the collapse of Chukotkan reindeer herding by dismantling social support systems).

But there is something both more complex and more subtle going on. The Soviet system was indeed flawed and the privatisation programme was indeed wrong-headed, but mechanisms that caused the collapse in Chukotka can be found in both (and can be reduced to neither). It would be a mistake to assume that all of the problems for Chukotkan reindeer herding started with privatisation, even if that is precisely when observers outside Russia began to notice them. In fact, by the 1980s, Chukotkan reindeer herding already seemed headed for a collapse that the privatisation programme merely expedited.

I argue, first, that it must be recognised that the Soviet system both systematically destroyed reindeer herding as it was practised in Chukotka prior to collectivisation – what is often called 'traditional' reindeer herding – while at the same time it reconstructed it in a new, Soviet form and then maintained its existence throughout the twentieth century. This new form was a reinvention in the image of the Soviet socialist economy, and by the 1980s any resemblance it bore to pre-collectivisation reindeer herding was for the most part superficial. In other words, that form of reindeer herding – a diversified subsistence economy that had been developed over many generations and which was rather beautifully self-sufficient – simply does not exist any more. What does exist is a thing created by the state, a thing which survives only by virtue of the extent to which it is propped up by that state.

Second, what privatisation triggered, then, was the collapse not of 'traditional' Chukotkan reindeer herding, but rather of the Soviet form with which it had been replaced. Since the Soviet form could only survive with state support, this is precisely why it is collapsing now that state support has been withdrawn, and why it had begun to deteriorate in the late Soviet period just as the entire Soviet economy was deteriorating. There is, therefore, no possibility of returning to a latent, pre-Soviet form of reindeer husbandry in Chukotka. However, there is the possibility of locally creating a new form of Chukotkan reindeer herding, one that is irrevocably affected by the Soviet legacy, but one which should be influenced, to whatever degree that local practitioners and managers desire, by notions of what ideal Chukotkan reindeer herding should be. There is already plenty of evidence that what is locally considered the ideal form is something known as 'traditional'.

In this chapter, I sketch out the historical contours of Chukotkan reindeer husbandry over the course of the twentieth century. I begin briefly with a snapshot of its prerevolutionary form, and then examine more carefully the Soviet project of collectivisation. Finally, I discuss the crisis that has befallen Chukotkan reindeer husbandry in the post-Soviet period, as well as the increasing importance of – and conflicting understandings of – the concept of 'traditional' reindeer herding. I argue here that this needs to be examined in a more critical way.

Prerevolutionary Chukotkan Reindeer Husbandry

It must first be clarified that the appellation 'Chukotkan reindeer husbandry' is a convenient and somewhat artificial construction, since historically there were variations in practices across Chukotka, blending into the practices of the bordering areas. Also, the word 'Chukotkan' by no means signifies that this is something practised only by the people known as the Chukchi; other reindeer-herding peoples include Evens, Chuvans, Koriaks and Yukagirs. Chukotka is also home to Yup'ik Eskimos, who reside primarily in the coastal areas of the Chukotka Peninsula and are associated with sea mammal hunting. There are also coastal Chukchis who historically engaged in sea mammal hunting from sedentary communities.

By contrast, the reindeer herding communities of inland Chukotka were historically nomadic. They practised a style of reindeer herd management that later came to be classified as the tundra type (Krupnik 1993: 88), in contrast to taiga reindeer herding. A key distinction between the two is that, with the tundra type, herds tend to be much larger and be used primarily as a source of food and raw materials, whereas with the taiga type, herds tend to be smaller and be used primarily for transportation (Ingold 1980). Chukotkan tundra reindeer herding, like Nenets reindeer herding, represents the extreme in terms of the size of the herds (1,500 head or more) and the reliance on nomadism, which was continual and could involve migrations of hundreds of kilometres. These long migrations enabled tundra reindeer herders to engage in long-standing trade relationships with coastal communities, in which reindeer skins, a key raw material of the tundra, used for the construction of dwellings, might be traded for seal oil, a key product of the coastal economy, used for heating those dwellings (Bogoras 1909).

Tundra reindeer herding communities tended to be loose affiliations of a handful of kin-related households who nomadised together and cooperated with other such neighbouring communities. Tundra reindeer herds were not held collectively or communally, but were owned and managed individually by families. Those who were poor in reindeer typically worked cooperatively with families who owned large herds, exchanging their labour for products of the herd (Krupnik 1993: 91). Pastures were typically shared cooperatively within these neighbouring communities, who worked out the seasonal rotation of pastures and the migration routes. These territories were flexible, so that in an emergency situation (such as a hard freeze of the tundra), neighbouring pastures could be temporarily used.

Due to strong local resistance, Tsarist government agents were never fully successful in imposing Tsarist administrative structures and policies on Chukotkan communities, although efforts had begun in the seventeenth century (Dmytryshyn et al. 1985). By the end of the nineteenth century, Russian and American traders had begun to make commercial inroads into Chukotkan communities, primarily among the coastal communities in the first instance. After the Bolshevik revolution, more concerted, and comparatively more successful, efforts were made to impose new Soviet administrative structures and policies on both coastal and inland communities, as well as to oust American traders, and by the 1920s the first efforts to collectivise Chukotkan reindeer herding were undertaken. What I will now turn to for the remainder of this paper is an overview of the process by which the Soviet state accomplished its thorough dismantling and restructuring of Chukotkan reindeer herding.

How Reindeer Herding was Collectivised in Chukotka

Collectivisation in Chukotka, as everywhere else in the Soviet Union, was driven by the ideology that the application of socialist economic principles would naturally improve any kind of economic activity, even something as alien to the average Russian as reindeer herding. In many ways, Chukotka's very remoteness and intractableness made it a particularly enticing challenge to the Soviet mission of bringing socialist enlightenment to every dark corner of the country. More so perhaps than other parts of the Russian North, Chukotka became a testing ground for experimental policies. The idea seemed to be that if it could be made to work in Chukotka – a place so far from the centre and so resistant to outside control – then it could surely be made to work anywhere. Therefore, Soviet planners were all the more zealous about pushing policies and strategies to the limit in Chukotka – the pace of collectivisation had to be faster, the *sovkhozy* (state farms) had to be bigger and more diversified, the reindeer herds had to be larger. Late Soviet-period scholars described the collectivisation process in Chukotka as 'marked experiments of social engineering aimed at destroying nomadic ways of life' (Pika 1999: 96).

Considering its remoteness from the centre, collectivisation started very early in Chukotka – efforts were already underway by the 1920s, and in 1929 Chukotka's first *sovkhoz* was established in the newly created village of Snezhnoe, along the Anadyr' River (where I conducted fieldwork in the 1990s). This was one of a handful of experimental stations set up in regions of the Far North to prepare the way for the establishment of *sovkhozy* as the primary organisational form of the reindeer economy (Druri 1989: 4). One of the founders, Ivan Druri, was sent directly from a post in the village of Lovozero, Murmansk, where the first experimental station for studying reindeer herding had already been set up in 1926. In Chukotka, Druri and his cohort began to travel among local reindeer herders. Writes Druri:

> We had to prove to the Chukchi that we wanted to organise a large Soviet enterprise and needed advice and help from local, experienced reindeer herders, and we also wanted to acquaint them with more rational modes and methods of doing reindeer husbandry practised by other nomads of our country. We lived and ate with the herders together in the *yaranga*.[1] We went out to the herd, helped where we could, and evenings we conversed with them, exchanged impressions, and talked about the work and life of reindeer breeders in the European North. (Druri 1989: 5–6)

The next *sovkhoz* in Chukotka would not appear until much later, and the Snezhnoe *sovkhoz* was a remarkable, although influential, exception. For the most part, collectivisation efforts began with the strategy of coaxing local reindeer herders to voluntarily form themselves into *artely* (cooperative associations) and the even more basic *tovarishchestva* ('comradeships'). This was seen by the organisers as a golden opportunity for the poor reindeer herder to strengthen his individual position through cooperation, and to facilitate more successful production – goals they assumed would be recognised by all as obviously desirable.

It bears mentioning that this was an urban-biased Bolshevik model of economic development that was simply being imported and imposed upon the Chukotkan population with little modification, much as it had been imposed on Russian peasants elsewhere (Shanin 1987). The model took as given that there was class differentiation between poor labourers and rich exploiters. The organisers of collectivisation simply sorted the available population into these categories and began to work from there to reproduce the process occurring elsewhere in the country, from urban factories to rural farms. The *artely* and *tovarishchestva* were seen as a preliminary phase, building blocks for the eventual construction of *kolkhozy* (collective farms).

Not surprisingly, the process was not without snags. At first, local Soviet administrators assumed that since Native society was at a level of 'primitive communism' (as Marxist theory had taught them), joining a collective enterprise, or *kolkhoz*, would be a relatively natural transition for the Native population, and they pushed hard for *artely* and *tovarishchestva* to join with one another and form *kolkhozy*. This over-eagerness led to some mistakes, as one late Soviet source openly admits:

> The underestimation or complete ignorance of private-ownership tendencies among the native population (especially the nomads), the socialisation of dogs, sleds, whaling boats, nets, weapons and even yarangi, all this provoked legitimate dissatisfaction. In many districts of Chukotka, this brought an increase in cases of armed resistance and attacks on the local activists, Party people and soviet workers. (Dikov 1989: 211)

Successful reindeer herders were alarmed to suddenly find their achievements reviled as the hoarding of *kulaki*,[2] and to hear themselves called 'class enemies'. In response, some herders fled from Chukotka altogether. Many of those who stayed behind slaughtered their deer to keep them from being handed over to the collective herd, a phenomenon that occurred across Russia.

The zealousness of local organisers was tempered when orders came down from above to recognise the existence of private deer ownership and to take a more liberal attitude towards it. Thus in the early 1930s *tovarishchestva* and *artely* were allowed to exist alongside the nascent *kolkhozy*. Later in that decade, conditions for membership in *kolkhozy* were mitigated; their voluntary nature was emphasised, and individuals were allowed to maintain some private ownership, including up to 600 of one's own reindeer (Dikov 1989: 213). By the end of the 1930s, the Soviet organisers in Chukotka had managed to organise a large number of *kolkhozy*.

In the 1940s, the emphasis shifted to the next step in the overall plan, which was the consolidation (*ukreplenie*) of the many small and dispersed collective enterprises into larger and more centralised units. The problem, from the point of view of the Soviet state, was that the Native people were scattered all about Chukotka in tiny camps and settlements, which was considered unthinkably inefficient for the purpose of developing and diversifying the tundra economy, not to mention incorporating the residents into the state system. In order to improve efficiency, larger groups of people would have to be brought together so that their efforts could be combined and their productive activities diversified (adding hunting and fishing to reindeer herding, or combining reindeer herding and sea mammal hunting) (Dikov 1989: 276). Consolidation was a burdensome process for the Native population across the Russian North; in many cases it meant closing down an entire village and moving its inhabitants into another village. On the coast, it meant ethnically distinct Chukchi and Yup'ik villages were forced to break down their historical distinctions and live as one community, defying the whole notion of dynamic cooperation between independent communities of tundra reindeer herders and coastal sea mammal hunters. In the tundra, it resulted in communities that were a multiethnic jumble.

A key goal of consolidation in the tundra was that of 'settling the nomads', i.e., tying down the nomadic reindeer herders to a central place, since Soviet planners considered such constant movement both unnatural and counterproductive. This goal of sedentarisation was by definition diametrically opposed to the very nature of Chukotkan reindeer herding, and the Soviet managers could not escape the fact that, with the large-scale tundra style of herding, someone had to be following the herd at all times. The solution they devised was a system known as the shift-work method. It entailed that brigades of reindeer herders would periodically be rotated out of the tundra to be resident in their village homes. Thus every Chukotkan 'nomad' in fact had an apartment in a village somewhere, where it was likely many of his family members lived, and his residence was officially registered in that village. Among both scholars and practitioners, many have blamed this policy in particular for sounding the death knell for Chukotkan reindeer herding.

The process of consolidation continued into the 1950s, and progressed alongside a complementary project, which was the reorganisation of all *kolkhozy* into *sovkhozy*. *Sovkhozy* differed from *kolkhozy* in that the property within them was not collectively held by the members, but was instead state-owned property. While the *kolkhozy* had members who theoretically set their own production quotas and paid themselves out of their own profits, the *sovkhoz* had employees who executed a plan handed down by the Ministry of Agriculture and were paid a salary and benefits by the state. Most significantly, this meant unequivocally that reindeer herding should no longer be considered a way of life developed by the herders themselves over generations, but rather a branch of the Soviet economy in which the herders were merely employees of the state.

As Humphrey (1998: 75, 93) has pointed out, the *sovkhoz* was considered to be a more advanced form of production than the *kolkhoz*, and should therefore in principle be reserved for areas of the Soviet Union that were generally more developed. In this sense, Chukotka (and perhaps the North in general) was truly an experimental region, a kind of laboratory of economic relations, for not only was the first *sovkhoz* already established by 1929, but by the 1980s all *kolkhozy* in Chukotka had been reorganised into *sovkhozy*, in contrast to southern areas of the Soviet Union, where *kolkhozy* remained in existence up until the time of privatisation.

The consolidation and reorganisation process continued through the 1960s and 1970s, which was a time of accelerated industrial development in Chukotka as well. By 1958, what had been nearly 100 small collectives was reduced to forty-one, and by 1968, the total number of collectivised enterprises was down to twenty-eight, a number that held steady until the beginning of the privatisation programme (see Table 9.1). At this point only three *kolkhozy* remained – the rest had been converted to *sovkhozy* (cf. Leont'ev 1977: 43, 162–71). In many cases what was now considered a single enterprise actually consisted of a headquarters with two or three branches that might be located a hundred kilometres away.

TABLE 9.1: The Collectivisation and Reorganisation of Chukotkan Reindeer Herding in Official Figures, 1920–2001[3]

Period	Number and type of enterprise	Reindeer headcount
1920s	before collectivisation	556,900 (100% private)
1930s	62 *kolkhozy*	427,400 (% decreasing)
1940s	76 *kolkhozy*	414,000 (20% private)
1950s	about 100 *kolkhozy*	408,422 (18% private)
1960s	41 *kolkhozy* and *sovkhozy*	571,400 (6% private)
1970s	28 *sovkhozy* and *kolkhozy*	580,500 (4% private)
1980s	28 *sovkhozy*	464,457 (4% private)
1990s	57 various privatised entities	148,980 (100% private)
2001	28 enterprises	85,947 (not all private)

Increasingly, these consolidated *sovkhozy* were treated as productive branches of the Soviet economy just like any other, made subject to the production quotas of centrally determined five-year plans. They also continued to be made subject to Soviet ideology. In the spirit of the Soviet Constitution's call to eliminate the material differences between city and village, as well as between the centre and the peripheral regions,[4] the concept of the factory work brigade was transposed even to the reindeer herders of the tundra and the sea mammal hunters of the Bering Sea coast. Ideally, the brigade should replace the family in both function and affection. In fact, throughout the Soviet period, one could still find family members working together in any given brigade; however, these brigades did not function as family-run herds, since they were closely integrated into the *sovkhoz* work plan, with constant contact with the *sovkhoz* headquarters maintained by radio and helicopter. Moreover, the brigade concept to some extent colonised the consciousness of reindeer herders, so that, to this day, groups of reindeer herders working in the tundra are referred to – and refer to themselves – by brigade number (and this seems consistent across the North – see Anderson 2000a). For example, in the village of Snezhnoe, a 'Brigade No. 4' existed in the tundra even though there were only two brigades left altogether in the *sovkhoz*.

The role of Natives in these collectivised operations shifted over the years. In the beginning, Natives were appointed as the veterinary specialists, accounting clerks and even chairmen of the *kolkhozy*. This occurred in part because of Lenin's policy of indigenisation (a kind of affirmative action policy that was meant to ensure that at least some positions were staffed by local non-Russians), but also in part simply because there were not enough trained Russians to cover all of the positions. Historical accounts take great pains to point out the high percentage of Natives staffing the *kolkhozy* in these early years as a great boon to the Native population and a credit to the Soviet system (Dikov 1989: 216). However, as more Russians arrived in Chukotka with desirable professional skills, the percentage of Natives in kolkhoz management positions declined, and the majority of Natives came to occupy a labouring class within the *sovkhozy*. This was more a by-product of systematic racism and the Soviet version of the 'glass ceiling' than a reflection of the real skill level of these Native managers – many Russians undoubtedly felt uncomfortable having Natives in superior positions.

Changes in the Practice of Reindeer Herding

All of the foregoing has sketched the process by which an attempt was made to force Chukotkan reindeer herding to conform to a Soviet organisational structure imposed from above. But what about herd management itself? Here also, a deep transformation was caused by Soviet management policies and practices. Throughout the Soviet North, the management of reindeer herding was centrally directed by mostly Russian and Ukrainian specialists in reindeer husbandry, some of whom had never had a first-hand encounter with a reindeer. They worked in regional agricultural, veterinary and land-use institutions (the ones who directed Chukotka were based in Magadan, capital of the province under which Chukotka was formerly subsumed). *Sovkhoz* directors then obliged local herding brigades to implement the herd management and pasturing plans of these specialists. As Krupnik points out, the Soviet state 'introduced large-scale, heavily subsidised, collectivised, and centrally planned reindeer industry, and it promoted this model with an iron fist' (Krupnik 2000a: 52). In this way, what was a very diverse set of reindeer husbandry systems and local herding practices, not only across the North but within any given region, became artificially uniform. In fact, although in Chukotka one can generalise about a large-scale tundra type of herding, historically there was variation in the size and composition of herds when comparing poor and wealthy herders. For example, the former placed more emphasis on transport deer (usually castrated bulls), while the latter emphasised breeding does, thus resulting in very different herd structures and managements tasks (Krupnik 1993: 103–4).

The key to achieving this uniformity was the creation of standardised reindeer husbandry manuals, produced in these regional institutions as well as in Moscow, which embodied Soviet policy and ideology regarding reindeer herding. These manuals covered techniques for feeding, breeding, veterinary care, slaughtering, and increasing meat productivity, and they stipulated the seasonal use of pastureland, indicating precisely where and when each brigade should move its herds. Some of the manuals were focused on particular regions of the North or particular ecological zones (for example taiga vs. tundra), while others were more comprehensive, covering the North as a whole.[5] These manuals began to appear as early as the 1930s, and new ones were produced continually throughout the Soviet period and beyond. One of the most recent manuals (Syrovatskii 2000) is remarkably similar to Soviet manuals except for one difference: the inclusion of a chapter titled '*Biznes-plan*' (business plan), which reads much like any socialist-era chapter on planning except for the mention of investors, bank loans and markets.

A classic example of such manuals is the 1950s sourcebook on reindeer herd management specifically in Chukotka, written by V. Ustinov, a zootechnical specialist with the Magadan land-use office. This early manual also gives clues about the motivation behind what came to be the accepted Soviet model of reindeer herding, i.e., one that pushed for the highest possible overall deer headcount as well as the highest possible average herd size.[6] Ustinov states in his introduction, 'the reindeer herders of Chukotka should seek to make their contribution in the business of increasing food products for the population and raw materials for light industry' (Ustinov 1956: 6). He then goes on to openly challenge the 'incorrect opinion of certain managers' who had gone before him that there was a certain optimal herd size that could not be surpassed (Ustinov 1956: 18), and later argues:

> Currently, when the task of obtaining high productivity and marketability of reindeer husbandry has been placed before the *kolkhozniki*, the old system of organizing herds cannot be suitable. Therefore in the last ten to fifteen years specialists and practitioners of northern reindeer husbandry... have worked out and applied in many places new forms of organizing the reindeer herd. (Ustinov 1956: 93)

Consequently he urges that the total headcount of reindeer in Chukotka – at that time just over 400,000 – could and should be increased to 683,000 head by 1960 (Ustinov 1956: 6). The highest headcount ever reached in Chukotka was just over 580,000 in 1970, a figure many consider to have been beyond Chukotka's carrying capacity. From there it began a slow and steady decline, until in the 1990s it crashed most precipitously (see Table 9.1).

The motivation behind all of these recommendations is made quite clear in Ustinov's arguments: increasing production was the highest priority of the Soviet economy, and nothing should stand in the way of achieving that priority. This encouraged managers of *sovkhozy* to ignore recommendations from land-use specialists about pasture carrying capacity and push for higher average herd size and a higher number of reindeer per *sovkhoz*. Chukotka had always been known for having the most reindeer of any region in the North, as well as the largest average herd size, even prior to collectivisation (Krupnik 1993: 100). Whether it was pushed by the planners above the Chukotkan managers, or was a result of the zealousness of those working within Chukotka, one way or another the expectation developed that Chukotka should always be pushing at the frontiers of the rural economy. The time-scale for consolidation of *kolkhozy* and conversion to *sovkhozy* was faster, the *sovkhozy* were larger and more diversified, the reindeer herds were bigger, and those herds were more productive in terms of tons of meat (Zabrodin 1979: 189–91).

In regard to the pasturing of deer, it should be pointed out that collectivisation had entailed a kind of socialist land enclosure of tundra pasturelands (Anderson 1991; Fondahl 1998: 72). When a *sovkhoz* was created, the regional land-use office carved out for it a strictly bounded territory, beyond which its reindeer herds were forbidden to range. Moreover, each brigade was also assigned a specific parcel within which it was obliged to stay, except for when the herd was driven to a central slaughtering point used by the entire *sovkhoz*. Further, all of the pastureland within each territory was studied and its characteristics determined, and on the basis of this information the seasonal rotation pattern of each herd was designed and written into the manuals. It was no longer important to have access to alternative pasture in the event of an emergency, since there were now specialists in the development of manufactured reindeer feed, which could be hauled in for the deer. What was far more important was to ensure that nothing that might interfere with the goals of high production was left to chance – nor to the skills of the herders themselves, in spite of their intimate knowledge of the pasturelands and their generations of experience working within it.

As for the Native reindeer herders, those generations of experience, having once been mined for information by experimental specialists like Druri, were cast aside as irrelevant to all but the ethnographers, who ironically described this as dying culture that needed to be collected and preserved. Young Native men, who in the past would have learned the practice of herd management by working alongside their male relatives of an earlier generation, were now taken out of the tundra altogether as children to attend boarding school, where reindeer husbandry was one of the subjects they learned from textbooks. Later, as young adults, some might be sent to agricultural vocational schools to be trained as specialists, studying zootechnical and veterinary aspects of reindeer herding alongside other subjects related to rural economy. In Snezhnoe I became acquainted with one of the *sovkhoz* veterinary specialists, who showed me his diploma from the Magadanskii *Sovkhoz*-Tekhnikum (technical school) in Ola, a small town near Magadan. He had passed exams in all the standard subjects printed on the diploma, such as the husbandry of cattle, sheep, pigs and poultry, as well as in reindeer husbandry, which was written in below.

This veterinary specialist also took me in to visit the headquarters of the former Snezhnoe *sovkhoz*, which by then already had the atmosphere of abandonment that became typical of post-Soviet *sovkhozy*. We stepped into the empty director's office, where he showed me a shelf lined with reindeer husbandry manuals. As we leafed through the pages and examined seasonal pasturage maps, he softly exclaimed, 'As if we needed to be told that. Our people long ago knew how to pasture reindeer – we didn't need a plan'. And yet the socialist plan came to replace the memory of those long-ago practices of his own ancestors. By the 1990s there was virtually no one left with first-hand memory of pre-collectivisation herding practices, especially in Snezhnoe, where nearly all of the male elders had died in quick succession from a variety of causes, including suicide. Any local knowledge about herding practices that these elders had managed to pass down could have been applied only in a piecemeal and subversive manner, since that knowledge would have been considered secondary to the authority of the manuals and the urgency of the socialist plan.

The Post-Soviet Reindeer Crisis and the 'Return to Tradition'

By the late 1980s there were already signs that Chukotkan reindeer herding was struggling to hold together, and the reindeer headcount had been declining since 1970. In 1991, then-president Boris Yeltsin began issuing a series of decrees requiring all state enterprises in Russia, including *sovkhozy*, to reorganise themselves into joint-stock companies or 'farming enterprises' (*fermerskoe khoziaistvo*) (Wegren 1994). In Chukotka, this reorganisation was accompanied by much rhetoric about Natives returning to 'traditional' forms of economy, regaining their ancestral lands, and finally becoming owners of their herds again. In practice, this is not how things turned out.

Some Chukotkan *sovkhozy* reorganised while managing to remain essentially intact, an approach that was generally preferred by the Russian directors. However, in some cases, individual brigades broke away from the main *sovkhoz* to form small independent enterprises, with the Native brigade leader now becoming the director of the enterprise (and in some cases Russians privatised their own herds, but these they typically slaughtered immediately, selling the meat for cash and moving on to other endeavours). In spite of the rhetoric about returning to the old ways of herding, these newly independent small herds did not resemble pre-collectivisation herds in any respect aside from their smaller size. Although in many cases they involved family members, one cannot say they were family-run, since they typically employed an outsider with some expertise in transport, accounting or marketing (and there were several cases of swindling and embezzlement by these outsiders). These enterprises now had to negotiate market conditions and become entirely self-sufficient, bargaining where they could for supplies, transportation, access to markets, etc., either with former *sovkhoz* directors or with independent traders. This proved more difficult than anyone had anticipated, and these new enterprises quickly began to fail.

Meanwhile, the rump *sovkhozy*[7] were also failing. In Chukotka, as elsewhere in Russia, the *sovkhoz* had not been merely an economic institution but a 'total social institution' (Humphrey 1995: 77). Virtually all of the social services of the Soviet village were administered by the *sovkhoz*. The process of privatisation entailed transferring most of these social services to other agencies either within the village or at higher administrative levels in the region. This left the *sovkhoz* with the seemingly simplified task of focusing on production. However, longstanding state subsidies that had kept the reindeer herding industry afloat during the Soviet period were no longer forthcoming, which in effect meant the instant disappearance of salaries for reindeer herders. By the time I arrived in the village of Snezhnoe in 1996, the herders had already gone several years without a pay-cheque, and were scrambling to find ways to get by. Their salaries were still calculated on paper, but so were their debts to the *sovkhoz* for advances of staple foods, and many herders were actually shown to have a negative balance in the *sovkhoz* accounts. Rather than being within the social safety net of the *sovkhoz*, they now had to request social services from other agencies, where *sovkhozniki* were now viewed as something like 'deadbeats'.

When I returned to Snezhnoe in 1998, there was a palpable sense of desperation in that village, and many herders had left the *sovkhoz* in search of more reliable employment. As was then common throughout the former Soviet Union, they expressed nostalgia for the former *sovkhoz*, which they said had kept them well supplied and comfortable when they lived in the tundra, and always paid a regular salary. They praised the free mobility afforded by the frequent helicopter flights to the village, saying now they felt trapped. Some told me that although they loved the tundra, they no longer considered it worthwhile to remain under such deteriorating conditions. It was cold there, life was hard, and they felt abandoned.

The health of Chukotkan reindeer herds suffered directly from the privatisation process, that is, from the removal of the Soviet system upon which herd health had been made dependent for so long. Since herders were abandoning the tundra, there were fewer people left behind to tend the herds, and quite often in Chukotka these were, ironically, younger men, with little experience in reindeer herding and fewer resources that would enable them to escape the tundra as they might wish. This was a generation that had grown up in boarding schools and were used to working under the enforcements of socialist work discipline; yet this enforcement had all but disappeared, and this was disorienting for the herders. Moreover, all of the resources they had taken for granted were now gone – scheduled air transportation, rifles and bullets, regular staple food deliveries, veterinary medicines, prosecution of poachers, guaranteed state markets, etc. Whether herders now worked in small privatised enterprises or in larger rump *sovkhozy*, all of these factors conspired to cause a cataclysmic crash in the reindeer herds. Deer were killed by wolves and bears, since there were no bullets to shoot them; herders were swindled out of deer by shrewd traders proffering alcohol; inexperienced or neglectful herders failed to follow basic management practices, resulting in losses of deer (for example, in Snezhnoe one year a brigade failed to drive the does across a river to join the bulls before the spring thaw, and thus there were no calves the following year). On top of all these factors of human error, in the 1990s there was an unprecedented increase in the number of wild caribou, and these wild herds began to draw away domestic deer as they migrated. Finally, the few herders who did keep their herds intact had great difficulty transporting slaughtered meat to urban centres where buyers could be found, and thus they had no profits with which to pay themselves and purchase supplies.

As a result of all of these factors, Chukotka, which throughout the Soviet period had been one of the most progressive reindeer-herding regions of the North, now represented the worst-case scenario – it had fallen the farthest of all. It became virtually impossible for reindeer herding to continue as it did either prior to collectivisation or during the Soviet period. The 'traditional' – i.e., pre-collectivisation – form was destroyed, and the Soviet form, which was dependent on government support, was no longer possible. Yet there have been frequent calls by Native activists and Native advocates for a 'return to traditional reindeer herding' in Chukotka, as well as throughout the Russian North, as exemplified in the following quote:

> The traditional foundation of nomadic herding life has adapted even to this [command-administrative] system. It exists in a latent and covert form hidden behind Soviet state farm documents... However, insofar as this foundation has not died, it provides excellent potential for the

revival of traditional herding among northern people, a tradition that is harmonious in its relationship with both nature and culture. (Elena Andreeva and Vladimir Leksin writing in Pika 1999: 96–97)

There is some very subtle conceptual slippage here. It is one thing to hold that one can today learn something about a form of reindeer herding that existed in the past (call it 'traditional') and employ that knowledge as an ideal, a model, in attempting to solve the current problems of reindeer herding. It is another thing altogether, and a fallacy, to contend that it is somehow possible to actually return to a past form. Time machines do not exist, after all.

What I have been arguing here is that, at least in Chukotka, 'traditional' reindeer herding – as a complex found prior to collectivisation – does not somehow exist in latent form, and the retreat of Soviet control over Chukotkan reindeer herding does not mean a 'traditional' form has automatically sprung up to replace it. This is not to imply that there was no resistance to Soviet domination – there was – nor is it to deny that there are subtle strategies by which traditions can be retained in spite of persecution, such as the 'bodily practices' described by Connerton (1989). Indeed, certain individual practices related to reindeer herding in Chukotka may be found scattered here and there in a form quite similar to the way they were practised in pre-Soviet times, such as the use of braided reindeer-skin lariats, the wearing of reindeer fur clothing, and even funerary practices involving the ritual slaughter of reindeer. My argument very specifically concerns that complex known as Chukotkan reindeer husbandry and the idea that, as a whole, it may be taken up again in its 'traditional' form.

I make the point not to reject outright the concept of 'tradition', nor to deny the possibility of everyday resistance, but because I think it is important, first of all, to demonstrate the process by which reindeer herding in Chukotka was transformed, and second, to be more precise in our concepts and terminology. To say that 'traditional' reindeer herding can be revived is to imply, however unintentionally, that indigenous peoples themselves are allowing it to lapse, perhaps by their own passivity and neglect, and that the intact, latent form could simply be taken up again, if only some people would muster up enough enthusiasm to do so. It should not be forgotten that cultural practices in relation to reindeer herding were forcibly altered or repressed. Therefore, the obstacles involved in practising reindeer herding today run deeper than mere willpower can overcome.

However, I would also argue that this is not so dismal an assessment as it might at first seem. It does not mean that indigenous people living in Chukotka are incapable of undertaking a reconstruction of their own local way of life in a manner that suits them today. And there is no reason why they should not do this under the banner of 'tradition', using their knowledge about past forms as a model for the future. Indeed, Native activists are studying written sources like Bogoras (1904–9) in order to 'remember' and reconstruct past practices. The very concept of 'traditional' as a desirable ideal might be seen as a new model of development offered as an alternative both to the Western neoliberal model as well as to the Soviet socialist model. In that sense it is a potentially powerful and useful paradigm for indigenous peoples, and constitutes a feasible strategy for working within the international system.[8]

However, scholars and practitioners alike need to be more precise and critical in their use of the term 'traditional'. It should not be used when what is meant in fact is 'historical' or 'former', nor should any conceptual confusion be allowed about the difference between looking to the past for models and actually returning to the past. One of the best programmematic statements of these ideas has been offered by the late ethnographer Aleksandr Pika, in his book *Neotraditionalism in the Russian North*. Writes Pika: 'We underscore yet again that the new "traditionalism" does not mean a return to the past. It is a forward-looking development, though one which attends to the specific nature of northern regions and peoples' (Pika 1999: 23). In Chukotka, it remains in question as to whether this enterprise will be successful in salvaging what is left of domestic reindeer herds in order to rebuild Chukotkan reindeer herding for the future.

Conclusion

Although the point of Soviet interference with the practice of reindeer herding was ostensibly to improve life for those practising reindeer herding, in fact the real point – and the result – was to make reindeer herding conform to the state's socialist identity. When the identity of that state was revolutionary (as in the 1920s), the collectivisation of reindeer herding was cast as a revolutionary process, with *kulaki* and class enemies to be rooted out. When that identity was industrial (as in the 1970s), the practice of reindeer herding was cast as a productive process, with production plans, quotas and competitive campaigns between brigades. When the state's identity became supposedly democratic and market-oriented, herders were ostensibly 'free' to choose what form of enterprise they wished to organise, and similarly 'free' to access the market. Some herders express nostalgia for Soviet-style reindeer herding, while others express a desire to 'return' to 'traditional' reindeer herding as it was practised by their ancestors. The challenge for all of them is finding the knowledge and the means to take up any form of reindeer herding in the future, regardless to what model it conforms.

Acknowledgements

This paper is based on the author's research in Chukotka, supported by a 1995–96 US Department of Education Fulbright-Hayes Doctoral Dissertation Research Abroad Fellowship and a 1995–96 Individual Advanced Research Opportunities Research Residency for Eurasian Russia from the International Research and Exchanges Board (IREX). Additional research was conducted in 1998 with support from the National Science Foundation's Arctic Social Sciences Program ('Regional Problems and Local Solutions in the Post-Soviet Transition: A Pilot Study to Assess the Problems Faced by Reindeer Herding Communities in the Chukotka Autonomous Okrug', award #OPP-9726308, PI Peter Schweitzer, Co-PIs P.A. Gray and M. Koskey). Research in 2000 and 2001 was supported by the Max Planck Institute for Social Anthropology under the direction of Chris Hann. This paper benefited from discussions with colleagues Deema Kaneff, Florian Stammler and John Ziker of the Max Planck Institute for Social Anthropology.

Notes

1 *Yaranga* is the reindeer-skin dwelling used by indigenous peoples of the Chukotka tundra, most closely associated with the Chukchi.

2 *Kulak* means literally 'fist', and it was the label that Bolshevik zealots gave to any peasant – or Native reindeer herder – who had accumulated more property than his fellow villagers, and was thus considered a mercantilist enemy of socialism.

3 Data from Dikov (1989), Leont'ev (n.d.), Ustinov (1956) and the Chukotka Department of Agriculture. Data found in the secondary sources is sometimes conflicting. 'Private' may include both deer held personally by members of *kolkhozy* and deer outside of the *kolkhoz* system altogether. The data presented here should be taken as snapshots in time, meant to give a comparative picture of how collectivisation progressed throughout the century.

4 Constitution of the USSR (1989 edition), Chapter 3, Article 19 (Verkhovnyi Sovet S.S.S.R. 1989).

5 See Syrovatskii (2000) for a comprehensive bibliography of these manuals.

6 The zealous pursuit of large herds applies primarily to what is called the *tovarnoe stado* or ('marketable herd'), which is the one from which deer were slaughtered. There was also the *plemennoe stado* ('pedigree herd'), kept primarily for breeding purposes, and these tended to be smaller.

7 Just as the term 'brigade' remained in common usage after privatisation, so did the term *sovkhoz* to refer to reindeer herding enterprises.

8 My ideas about 'tradition' in this context have been strongly influenced by discussions with my colleague Florian Stammler at the Max Planck Institute for Social Anthropology.

10 A Genealogy of the Concept of 'Wanton Slaughter' in Canadian Wildlife Biology

Craig Campbell

> I think the natives must be told, 'the good old days' are gone forever. The continued unregulated killing of migratory barren-ground caribou will destroy the great herds within our life time. If the native peoples do not soon exercise restraint and limit their take of caribou; if they pursue in not accepting the facts that have been put before them; rebuilding of the caribou herds, if it is possible, will take decades. Future generations of Canadian natives may never know the pleasure of the caribou hunt!
>
> F.L. Miller 1983: 173

There is a common policy in most northern regions today, from Canada to Siberia, that local people and scientists must work together in order to manage Arctic landscapes. In the 1990s and the start of the twenty-first century, 'co-management' has been a key word in the relationship between scientists and indigenous hunters of caribou. It is no secret that the political and in many cases the legal imperatives that encourage biologists and traditional hunters to work together makes an uneasy alliance. Several key articles (Bergerud 1988; Osherenko 1988; Freeman 1989) as well as chapters in this collection (Usher, Nagy, Sejersen) record difficulties in the purpose, paradigms and methods of this collaboration. To summarise an often-quoted critical reflection on the difference between scientific and traditional approaches to knowledge (Thomas and Schaefer 1991), scientific investigation is distinguished for being quantitative, predictable and based on verified information, while those using traditional knowledge favour qualitative, ethical or hearsay information.

In this chapter I wish to question the widely held assumption that caribou biology has special access to truths about caribou behaviour and population data and that it is qualitatively superior to the knowledge of other peoples. I conduct this analysis not so much to accuse individuals of 'bad' science but instead to bring to the forefront the much more humble and yet much more human way that wisdom is formed. It is my hope that this will be useful to help temper scientists' rhetorical claims of singular authority to speak for caribou. I will conduct a bibliographic investigation into the history of scientifically cited cases of 'wanton slaughter' to demonstrate that, at least in so far as positivist biologists select topics for research, they unquestioningly accepted assumptions in the written literature. I do not conduct this investigation to denigrate wildlife biology but instead to demonstrate that ethical and political questions dominate the agenda of both 'users' and 'scientists' when it comes to their choosing what is worthy to know and to study. Through a better understanding of how scientific truth is constructed, it might be possible for communities and academies to come to a reasoned understanding of the tasks that they can share together.

For reasons of space, I have chosen to analyse one key paper (Miller 1983) as a work representing the conventional wisdom of the 1980s. However, I will quote from an extensive literature (including Banfield 1954; Kelsall 1968; Parker 1972; Theberge 1981; Nudds 1988; Bergerud 1988) to demonstrate the pervasive circulation of the ideas cited. By deconstructing a single article, I draw attention to some important details which were sewn into the fabric of caribou management and which threaten to persist if they remain unchallenged. The technique that I use is to portray a genealogy of references that depicts the process whereby opinion creeps into scientific work, framing its agenda. Within anthropology, similar genealogical analyses have been conducted to uncover the 'ivory-tower myth' of Yukagir writing systems (DeFrancis 1989) and myths of aboriginal people being 'original ecologists' (Krech 1999).

I have constructed a genealogy to situate this key article in a larger body of published research concerning the caribou of northern central Canada (Figure 10.1). In the genealogy, I take a number of popular and canonical works on caribou biology and management from central Canada and draw out the networks of cited references. The connections drawn out in this graphic show how several authors rely on a single report to denigrate indigenous hunters and to legitimise their own authority. By charting the references I comment on how they are related to the discussion of wildlife management and the manner in which they are used and sometimes misused to support an argument. Such a genealogy is meant to uncover biases and assumptions embedded in canonical works, studies that have themselves become entrenched in the literature through frequent citation.

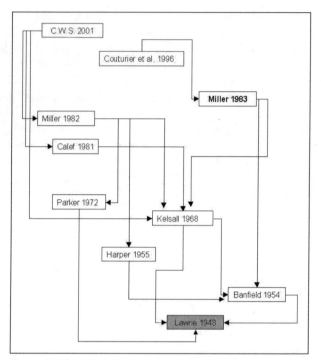

Figure 10.1 Genealogical chart of how certain authors build arguments by citing misleading information about aboriginal hunting practices.

Because canonical works continue to be cited in the twenty-first century without critical treatment of, or even reference to, the indictments within, there is a danger that partial knowledge framed as truths will continue to be reproduced as biases among wildlife managers. For example, in the website of the Canadian Wildlife Service (2001), three out of the four recommended readings in the 'barren-ground caribou' section contain explicit and highly contentious characterisations of indigenous overhunting.[1] While my treatment of Frank Miller's (1983) paper will perhaps seem harsh, I feel, nonetheless, that the weight of the accusations brought forward by Miller, in conjunction with the authority he held as a government-mandated scientist, more than justifies my position. Finally, the potential ramifications for indigenous northerners – who rely on wild meat for both dietary health and cultural wellbeing – resulting from policy based on the biased recommendations of well-meaning wildlife managers, is troubling in the extreme.

In reviewing Miller's article I will address three significant assumptions that underpin caribou management and undermine the scientific authority claimed by caribou biologists. I characterise these assumptions as 'the scientific veil', 'primitive hunters and their fall from Eden' and 'neocolonialism and the primacy of caribou biology'.

Neocolonialism and the Primacy of Caribou Biology

'Restricted caribou harvest or welfare – northern native's dilemma' (Miller 1983) was written for the journal *Acta Zoologica Fennica* as 'an opinion paper on the apparent impact that hunting by native peoples is having on large herds of migratory barren-ground caribou...' (ibid.: 171). This special issue of the journal was devoted to reproducing the lectures given at the 3rd Meeting of the International Caribou and Ungulate Biologists – a key event for circulating ideas and techniques in wildlife biology, worldwide. Miller, a biologist working for the Canadian Wildlife Service (CWS), had been studying caribou in central Canada since at least the late 1960s. In brief, the argument put forward in the paper is that science-based wildlife management must be accepted by indigenous hunters if Barren Ground caribou are to be preserved from extinction. The author documents an apparent case of population decline among the caribou of northern Manitoba and attempts to show that 'native' hunters are responsible for this decline, co-management regimes are ineffective and that a self-imposed moratorium on hunting is the only solution to prevent the imminent demise of the caribou.

The anthropologist Robert Paine, in his 1977 book *The White Arctic: Anthropological Essays on Tutelage and Ethnicity*, notes cases of a paternalistic syndrome among non-indigenous scientists working in the North. Miller's article is typical of wildlife biologists of his generation who were burdened with a sense of responsibility over the northern menagerie, demonstrating an attitude of 'father knows best' (or in this case 'biologist knows best'). At the core of Miller's article and others like it is a profound distrust of 'users' (hunters), embedded assumptions about the ownership of natural resources, a strong belief in the truth-claims of wildlife biology, veiled cultural assumptions that parade as scientific truths and the lack either of capacity to understand or interest in indigenous hunting systems. In 'Restricted caribou harvest', Miller obviously feels powerless to halt the decline of the caribou he had studied for many years. In his desperation the neocolonial and managerialist underpinnings of caribou biology and management become apparent. Milton Freeman (1989), Harvey Feit (1998) and Fikret Berkes (1999) all describe these as paternalisms that were directly linked to the structure of resource management regimes of the Canadian federal government. Peter Collings, an anthropologist who worked with Inuit hunters in the Northwest Territories, cites Miller when he cautions against 'the preconceived notions about Natives and Native hunting that non-Natives, including some wildlife managers, may bring with them into the field' (1997b: 42). In terms of biological science, Freeman has been at the forefront of calls for the history of an honest wildlife biology. He writes that 'scientists have been trained to believe that their approach to understanding nature and their resulting expertise, represent the best available approximation to 'the truth' and should therefore form the basis of rational decision-making in management matters' (1989: 95).

Beginning in the postwar years when the Canadian Wildlife Service came into being, widely publicised indictments of waste and over-hunting of caribou were launched against indigenous hunters. Many of these accusations were poorly supported and appear to have been pro-moted by an uneasy wildlife management regime witnessing apparent population crashes of caribou. In 1956 A.W.F. Banfield, then chief mammologist for the Canadian Wildlife Service, wrote the most visible and graphic condemnation of indigenous hunting practices to date. In *Beaver*, a popular national monthly magazine, Banfield described how modern scientists had uncovered a critical decline in caribou popula-tions throughout the Canadian North. The decline was linked, in his opinion, to 'orgies of killing' by Eskimos and Indians. The argument was illustrated with a large photograph of a very successful local hunt for caribou (Figure 10.2), misconstrued in the text as 'abandoned car-casses' and as 'a scene of carnage'.[2] The place of these accusations as conventional wisdom (couched in the rhetoric of science) within the canon of seminal works in wildlife biology, as I show in this paper, is deeply entrenched and threatens to persist as internalised wisdom among generations of future wildlife managers.

In his 1983 article, Miller claims that historical instances of caribou decline have been precipitated by the uncontrolled indigenous hunt for caribou. He writes that 'Caribou biologists have had, at least, one consensus throughout the 33 years of caribou studies in Canada. That is, the human [read indigenous] kill of migratory Barren Ground cari-bou is the primary cause of decline in sizes of herds' (1983: 171). While Miller cites no one to support this statement, there are several ways of interpreting this alleged consensus of caribou biologists. In the appar-ently dire circumstances outlined by the author and identified as the

A scene of carnage where swimming caribou have been speared from canoes. Each year thousands of carcasses are thus abandoned.

THE CARIBOU CRISIS   By A. W. F. Banfield

Figure 10.2 'The Caribou Crisis'. The title page and photograph from Banfield's (1954) influential article allegedly portraying a case of 'wanton slaughter'.

factual outcome of research conducted by 'professional caribou biologists' (174) the imminent demise of the migratory caribou can only be curtailed by an indigenous, self-imposed, restricted caribou harvest. In Miller's view, intensive wildlife biology discovered a pattern of indigenous overkill (171). If indigenous hunters do not accept this 'fact' (172), Miller states, then they will undoubtedly destroy their caribou resource. The apparent intention behind this statement is that 'native' hunters will surely not want their caribou to disappear, they will concede to the authority of the science-based wildlife managers and will behave themselves. In laying out a history of unrestrained zeal in the caribou hunt, indigenous hunters are presented as no less than savages or primitives (that is, uncivilised) lacking in bodily control.

As was noted in the introduction, modern wildlife biology is a young science that developed in a period following the Second World War. In Canada, caribou science began in the Northwest Territories in the 1950s with A.W.F. Banfield's 'Preliminary investigation of the barren-ground caribou' (1954). Banfield, himself, writes that while caribou were first described by Europeans in the eighteenth and nineteenth centuries, 'little more was added for more than a century' (1956: 3). That is, until Banfield did his population surveys. Many articles on caribou management in the Canadian North reference publications from the infancy of caribou biology. In doing so they risk using biased and generally unchallenged information about indigenous hunters of the Barren Grounds. Since northern wildlife, like caribou, were never methodically studied or even counted by urban scientists prior to the 1950s, most biologists rely upon written accounts from fur traders or police to give depth to their environmental history. Banfield's study like many of the earliest sources that are uncritically used by current wildlife biologists, derives from these early sources which are often used to belittle indigenous hunting systems. Furthermore, as Robert Brightman (1993) notes, because of the frequency of plagiarism 'by the early chroniclers of Hudson Bay', such accounts 'need not all be taken as independent attestations' (288). Banfield (1954) and Kelsall (1968), two of the more frequently cited caribou scientists, have devoted significant portions of their writing to justifying state management of caribou by demonstrating how indigenous peoples are a danger to the integrity of the nation's caribou heritage.

In the defamation of indigenous hunters, or 'usergroups', as they have come to be known, the position of state management finds its legitimisation. Warning against unbound scientific rhetoric, Milton Freeman (1985) writes that wildlife scientists and managers can often be characterised by 'the conviction that the scientific approach to game management is superior to systems espoused by other groups of people', including indigenous peoples who are assumed to 'have neither the knowledge nor the institutional means of managing natural resources' (see also Banfield 1954; Kelsall 1968; 266; Theberge 1981; Miller 1983). Aside from disciplinary arrogance, there is a dangerous tendency for opinion to lodge itself in scientific research and become established as paradigm (Bradshaw and Hebert 1996: 229). In a recent article, Harvey Feit critically appraises 'the goals for action [of wildlife management, which] do not flow directly from the encounter with wildlife, or simply from abstract conceptual developments in science, but from social and historical ideas about what is best for wildlife and for some specific groups of people' (Feit 1998: 133). The claims, then, that caribou biology and management has privileged access, through rationality and scientific objectivity, to truths about caribou are politically and methodologically suspect. A common position occupied by biologists is suitably typified by John Theberge, who wrote that 'Early man [i.e. Native Americans] often failed to conserve... because he lacked the two prerequisites for conservation of resources: perception of the danger of over-exploitation and an option to do something about it' (1981: 281).

Primitive Hunters vs. Primitive Caribou: The Fall from Eden

> Many people are under the impression that the primitive Indian was a dedicated conservationist. Nothing could be further from the truth. The primitive Indian was limited to weapons which would not allow him to kill much more than he actually required. With the introduction of firearms to the northern Indian during the 18th century, the numbers of caribou killed increased substantially.
>
> Parker 1972: 74

Indigenous blood-lust and a zeal for killing are noted by a number of early European explorers in the Barren Grounds. Banfield (1954), Kelsall (1968), Parker (1972) and Miller (1983) all cite Lawrie (1948) among others (Hearne 1968 [1795], Russell 1898, MacFarlane 1905) as part of the 'numerous literature references to appalling slaughters and to the disappearance of caribou from accustomed ranges' (Kelsall 1968: 146). MacFarlane, for instance, at the turn of the century, wrote that '[t]he northern Indians were accustomed ... to slaughter thousands of reindeer annually, chiefly for the skins and tongues and too often from the sheer love of killing' (1905: 680). Lawrie similarly reports: 'the Ihalmiut, with no abstract ideas of conservation to restrain him and with apparently limitless numbers of caribou migrating through his country, takes what he can get with reckless prodigality' (1948: 25). While Lawrie did not extrapolate his claim, Miller (1983) and Kelsall (1968) were so moved by this passage that they would conflate Ihalmiut hunting practices with those of all other indigenous hunters. The failure to acknowledge many different nations of indigenous peoples over the grossly general term 'Indians' makes a serious logical error and is consistent with my characterisation of a wildlife service in denial of its own biases.

It is clear through my reading that these early accusations and indictments were decontextualised observations that have been removed from their geographical locale and used to support a campaign to defame indigenous hunting practices. The majority of the men who noted the wastage were not capable of conversing with their indigenous hosts, they did not stay with or hunt with these people for long periods of time and they were evidently so clouded by ethnocentric biases that they could scarcely have interpreted positive signs of indigenous management if it was plainly laid before them. James Isham (1949: 81), for example, wrote in the late eighteenth century:

> I have found frequently Indians to kill some scores of Deer and take only the tongues and heads and let the body or carcass go a Drift with the tide, therefore I think it's no wonder that godalmighty shou'd fix his Judgemen't upon these Vile Reaches and occasion their being starved...

Harper writes, similarly, that one 'band of Eskimos is said to have once slaughtered 500 animals, half of them in the river, where they did not even bother to pull them out; they had killed for the sheer delight of killing rather than for utilization' (1955: 49). On the following page the author then provides an explanation for the practice of river storage: 'At the Windy River post, in the latter part of summer, portions of caribou bodies are laced on the river not merely for refrigeration, but for protection from blow flies' (ibid.: 51). Harper never reconciles these two conflicting descriptions of caribou hunting.

Figure 10.3 The shameful waste of barren-ground caribou (Canadian Wildlife Service no date [1950s]).

In one of the most recent descriptions of caribou hunting *en masse*, Banfield (1954) writes of hunts that took place on river crossings: 'During these occasions the hunters became greatly excited and great slaughter of caribou took place' (Banfield 1954: 216). Clearly restraint and bodily control are primary metaphors used by critical biologists and explorers to explain the behaviour of apparently uncivilised natives (Figure 10.3). In one of the books cited through the Canadian Wildlife Service's website (2001), Calef writes that,

> [a]mong the native peoples the age-old hunting instincts have not died. It is almost impossible...to understand how much they equate the abundant killing of caribou with survival and happiness. Because the modern technological society has come to the North so suddenly, the instincts evolved over thousands of years have not had time to adjust...To now be told that taking this abundance will destroy the animals they cherish, the animals that have always sustained them, is more than they can bear. (Calef 1981: 165–6)

One central issue that arises from these early reports is the amateur ethnographical observations that were tainted by deeply held hunting aesthetics and ethnic stereotypes. John Sandlos writes of the idealised conceptions of the northern wilderness among Victorian-era Euro-Canadians, who held a 'strong attachment to a hunting code of ethics that abhorred the wanton slaughter of the abattoir and favoured the more sporting pursuit of a nimble quarry' (2001: 10). The idea of wastage by indigenous hunters is constantly remarked upon by naturalists and biologists working in the Arctic. According to one biologist, the meat of caribou, which are charismatic animals in Euro-American thought, should only be consumed by humans and not used 'ingloriously as dog feed' or abandoned on the land (Harper 1955: 48). It is further written by Harper that cached meat is more often than not enjoyed by 'the beasts of the field' (bears, wolves, foxes, weasels, wolverines, lemmings, etc.) who 'help themselves to the free feast' (ibid.). The hunting aesthetic of the Western biologists and managers seems to focus romantically upon a relationship between the hunter and the prey whereby the fruits of the hunter's catch are the sole property of the hunter, who must tragically kill the noble beast and use its body as intensively, or as completely as possible.

The anthropologist Robert Brightman provides the most compelling ethnography of indigenous hunting in northern Manitoba and shows that overkill is a far more contentious and complicated issue than it is made out to be by wildlife biologists. He writes that for the Rock Cree, '[t]here existed no conception of "waste" attached to the material bod-

ies of animals...' (1993: 283) whose souls, if treated respectfully, would
be recycled as more caribou. It is worth considering his position at
length:

> 'Waste' and 'overkill' could occur in this cosmos but only as events of rit-
> ual omission. The numbers of animals in the world, their distribution on
> the landscape and their accessibility to hunters were all conceived to be
> determined by the wishes of the immortal animals themselves or by
> those of their deific 'owners'. The nominal death of an animal was only
> one moment in a cycle: animals live in the bush, are killed by hunters,
> persist as souls after their bodies are eaten, and return again to the world
> through birth or spontaneous regeneration...The caribou left 'either to
> rot, or be devoured by the wild Beasts' (Ellis 1967 [1748]: 85) remained
> in the bush, there naturally to return to life...It was only the *ahcāk*, or
> 'soul,' of the animal that could be 'wasted' by neglecting the practices
> that facilitated its rebirth. It was in the eat-all feast and in the mortuary
> deposition of bones in water and on trees that Crees and other Algo-
> nquians 'managed' animals. (Brightman 1993: 288)

The issue of caribou meat being wasted on non-humans can be
addressed through the widely remarked-upon practice by northern
hunting peoples of considering animals to exist within the social world
as persons (Ingold 1996). Cree and Chippewyan hunters that I have met
have claimed that the carcass of an animal must be treated with respect
– though, as Brightman shows, this does not necessarily mean that
selectively using the meat from an animal is disrespectful. The language of respect should not be misconstrued to be an affirmation of
Western hunting ethics. In one of Lawrie's accounts of a hunt, some
wounded caribou escaped into the bush. While the Ihalmiut hunters
were uninterested in pursuing them, Lawrie insisted that they track
down the animals so that they would not be wasted (1948: 28). If the
crippled caribou were to die in the forest, as Harper noted, it would be
eaten by the other animals. Given that these other creatures exist, for
northern hunters, within the social world with humans, rather than
existing on a lower echelon of a hierarchy, what naturalists and biologists conceive of as waste may be understood as sharing, or something
altogether different. Furthermore, allowing crippled animals to be
eaten by scavengers may have been an intentional act by indigenous
hunters to 'feed' these other creatures. It stands to reason that contributing to the diet of the wolves, foxes and weasels, that you will later
hunt could be motivated by an active interest in the productivity of the
bush economy. At any rate, the point of this article is not to answer the
question of what hunters of the central Arctic were doing but rather to

show the cultural underpinnings of policy recommendations by wildlife biologists.

Another troubling issue in Miller's paper is the notion of primitive numbers of caribou. Following Kelsall (1968: 144–5), Miller writes that '[p]rimitive numbers of migratory barren-ground caribou persisted in northern Canada until the advent of firearms shortly after 1700' (1983: 171). Although theorising a difference between primitive and modern caribou is an attempt to account for changing patterns of use, it presumes an implausible primeval state of sanctuary whereby the supposed natural equilibrium that governed the primitive world was upset by the advent of modern technology. The idea of primitive numbers suggests an assumption, deeply entrenched in Western thought and particularly in wildlife management, that humans are distinct from the supposedly natural world that surrounds us (Ingold 1992). Whether we understand the human state as one of progressive, evolutionary ascent from simple to complex beings or as divinely empowered with an hegemony over all worldly things, or some mix of the two, we have at least begun to question the culturally based roots for contemporary caribou management.

Feit explains the problem of the cultural underpinnings of wildlife management by stating that what is 'implicit in much of the discussion [about wildlife management] is that resource users do not and cannot consider the interests of the exploited wildlife and therefore a specialized and disinterested agency is needed' (1998: 131; for an example of the literature that may contribute to managers' antagonism towards indigenous peoples see Hardin 1968 and Kay 1994). Furthermore and perhaps more importantly, managers have shown either little capacity to understand or any interest in the 'frequent active stewardship of wildlife' by indigenous peoples (Feit 1998: 132).[3] The manipulation or management of boreal landscapes is in part documented by Henry T. Lewis in *A Time for Burning* (1982). Lewis recognises that the way that Dene and Cree people in northern Alberta altered their geography through fire to promote grazing pastures is just one example of indigenous manipulation of 'natural' spaces (Lewis 1982). By extension, the supposedly primitive numbers of caribou, untouched by the technology of civilisation, may have been, at least in part, actively manipulated by the 'native users'.

The idea that 'overkill became excessive with the efficient weaponry of the white man' (Theberge 1981) is paradigmatically applied to indigenous hunters of the Canadian sub-Arctic and Arctic: the entirety of so-called 'native' hunting is understood to have been primitive, inefficient, and irrational. The implication of these indictments is clear while the evidence remains suspect. Without reference to any particular source Miller writes that 'the so-called harmony [that existed between the caribou and the Inuit and Indian hunters]... was imposed by the caribou's continuous movements; the native's relative lack of mobility; and the native's poor weaponry' (1983: 173). According to the rhetoric of the wildlife managers, once it has been established that there was never any indigenous management, just primitive technology, it follows that 'natives' have abandoned the Garden of Eden through their use of modern technologies. Quoting Harper's 1955 study of the Barren Ground caribou at length, Miller apparently agrees with Harper's revealing description of the 'caribou resource' whose 'Garden-of-Eden trustfulness in the presence of man...[makes it] more worthy of being cherished and safeguarded in its natural haunts for the benefit and enjoyment of future generations' (174).[4] This inclusion is a good example of the Euro-American cultural assumptions that back the supposedly objective science practised by wildlife biologists. The charismatic nature of caribou evidently places them higher on a hierarchy of worth than fallen 'Indians'.

In accordance with the situation of caribou crisis, Miller writes that '[o]nly through intensive, biologically sound management with full native *cooperation* will caribou remain an item in the diets of northern Canadians' (1983: 171; emphasis mine). We could exchange cooperation with *submission* and have a more accurate portrayal of Miller's position. It is apparent that he feels caribou managers are being forced to accommodate the 'natives' due to federal and provincial governments that were showing an increasing interest in compromising the authority of state wildlife managers. Miller rejects the possibility of co-management, discounting it as an unmanageable fantasy and one that will ultimately result in the destruction of the caribou. In Miller's paper, F.J. Tester (1981) is criticised for having 'rose-tinted glasses' (1983: 194) when he calls for greater effort on the part of managers to understand indigenous hunters. Tester (1981: 195) notes a marked contention between remote (non-local) biologists and local (in this case indigenous) users and recommends:

> the acquisition of greater knowledge by biologists and wildlife managers of the social, psychological and cultural factors of relevance to the populations their efforts should ideally serve...There is a need for serious and in-depth training related to human needs, values, the act of valuing, differences across cultures, the importance of and subtleties of meaningful communication and the significance of participation.

Miller's conditions, on the other hand, stress the need for hunters to conform to wildlife biology and management. His interest in management is unbending, rejecting Tester's cautionary warning on the importance of 'participatory democracy' (ibid.). The interests of the Canadian state are reasserted by Miller in his statement that the 'well-being of those [Barren Ground caribou] herds should be the concern of all Canadians' (174). Miller is apparently doubtful of the possibility of '[l]ocal self-control and support of written law for native management institutions' to be successful as he warns that we 'should not run the risk of well meaning ignorance leading to calamity for the caribou' (174). His paper is a clear statement of the necessary primacy of the 'existing established wildlife management and research agencies and professional caribou biologists' (174). While specifically focusing his research on the Beverly and Kaminuriak herds, Miller generalises 'about the probable impact on all migratory barren-ground caribou' (172). One problematic case is thereby generalised as exemplary of all regions in the Canadian Arctic and sub-Arctic regions.

The Scientific Veil

The most damaging and offensive work by wildlife mangers results from an unwillingness to be self-reflexive, and an unbending, fundamentalist belief in the primacy of their discipline. In another section of Miller's paper, the author attempts to show how the 'natives" claim that they need five to seven caribou per person per year would deplete the herd by 70 to 100 percent. He thus dismisses the 'native's so-called needs...[as mere] desires' (172). Miller then goes on to outline a theoretically sustainable harvest limit which ends up looking like one caribou for seven to ten 'natives' per year (172). In writing that the natives paraded their naïve and selfish desires as needs, Miller conjecturally builds his case that indigenous hunters are the enemy of caribou. Furthermore, the word 'desire' is consistent with my characterisation of the entrenched neocolonial position, which imagines indigenous hunters as savages that lack control over their bodily urges. Thus we see the use of words like 'self-control' and 'restraint': terms that have a central place in the ideological underpinnings of Euro-American hunting.

A clear example of the misuse of sources is seen in Miller's treatment of caribou population estimations. To make an argument about the upper limits of caribou productivity in Arctic ecosystems, Miller cites an article where the authors attempt to calculate the growing human population to compare results with their measurements of the ecological productivity of the land (Fuller and Hubert 1981).[5] It makes for a simple input-output system that may have little correlation with reality. In fact, Fuller and Hubert concede that 'in the absence of historical records it is even more hazardous to estimate future animal populations than it is to project human populations' (1981: 15). The authors write that their calculations for predicting future caribou populations are based on the 'well-researched Coats Island herd' (17). The findings from this study are then extrapolated to predict mainland herd productivity. In Frank Miller's article, he makes reference to Fuller and Hubert's estimates for the indigenous population of the Northwest Territories in the year 2000 but he does not cite their warnings about the hazards of projecting human populations. Furthermore, extrapolating from the Coats Island herd may be a flawed theoretical projection. As one anthropologist noted to me, this herd is an inappropriate model for mainland Barren Ground caribou because Coats Island provides a very restricted range for an expanding herd, there are no wolves, the caribou are not stressed by undertaking two huge animal migrations to the boreal forest and back and, finally, there is very limited human predation as few people have large enough boats to make the journey.[6]

In another example, Miller's article was recently cited by Couturier et al. (1996) to support a theoretical estimate advanced by Miller for the crippling loss of caribou. 'Crippling loss' is a term used by wildlife biologists to refer to animals, wounded by hunters, that are not captured or 'harvested' and subsequently die in the wilderness and, according to Western hunting aesthetics, are thought to be wasted. Ironically, Miller provides no support for an estimated 20 percent loss of caribou due to being crippled in the hunt, thus putting a major component of the Couturier et al. study in question. The irony is more troubling than amusing because Couturier et al. are citing a paper that presents an aggressive and highly anecdotal indictment of indigenous hunters. One can only guess when Couturier et al.'s article will be reproduced to further support arguments and claims about 'crippling loss'.

Biology in the service of the nation's caribou has aroused 'deep suspicions about the relevance of science and ... a legacy of doubt about the ability of science to work in the interest of anybody other than scientists and southern institutions' (Meredith 1983: 102). Doug Urquhart writes that 'understanding caribou population dynamics has proven to be a Pandora's box of statistics and complexities, whose solutions always seem to move just beyond the capacity of available data to resolve them. This in turn, sets researchers on an endless quest for more and better information. Ultimately, however, such information may be unobtainable at any reasonable cost' (Urquhart 1996: 266). Partial data and limited budgets have, perhaps, precipitated hasty policy recommendations from wildlife biologists. Despite acknowledging his own questionable harvest data, for example, Miller felt confident in stating that 'native kill is the principle cause, if not the sole cause [for the decline of herd size]' (173).

Miller's invective paper was a response to the perceived decline of the Kaminuriak herd in the early 1980s. In hindsight, it seems that moving populations of caribou, rather than overhunting, were found to be responsible for what caribou biologists thought was a population crash. In Bergerud's 1988 article, the author notes the occurrence of major movements between caribou herds. He details this with an example of a great movement out of the Kaminuriak Herd, bordering on the Beverley herd, which in three years dropped by 110,000 animals (out of 149,000 in 1955). He writes that '[t]wenty-five years later the Kaminuriak Herd showed another unexpected change in numbers; it increased from 39,000 in 1980 to 180,000 in 1982' (Bergerud 1988: 108). In a similar analogy, Collings notes that Inuit hunters in the Holman area of the Northwest Territories cited out-migrations of caribou as a cause for the decline in numbers – challenging the accusation that their hunting practices were the primary cause of an apparent population crash (1997: 50). Both of these cases can be used to suggest that Miller's caribou crisis was a shift in allegiance of caribou from one herd to another.

Conclusion

While the veracity of the widely reported indigenous wastage of caribou is difficult to substantiate and impossible to universally apply to all indigenous hunters, it is clearly taken by caribou managers as an indication that some, if not all, indigenous hunters are inherently wasteful in their hunt. Banfield (1954: 59–60), Kelsall (1968) and Miller (1983) all report hearing of great numbers of caribou that are wasted by indigenous hunters. However, as my bibliographic genealogy shows, these authors quote the same report (Lawrie 1948) as evidence of indigenous waste in hunting. Lawrie's report, on closer inspection, offers poor evidence that all 'Indians' were inherently wasteful hunters. On examination, a number of the most frequently cited sources from the turn of the century and earlier are equally problematic. In Miller's article, we see how the weight of many accusations is taken as an indictment that demands intensive state-mandated management regimes. It is not obvious in my reading of the literature that unused caribou are a sign of indigenous ignorance and savagery, as the above authors suggest. I am more interested in the differing aesthetics and ethics involved in hunting. For example, it might be at least as productive to ask what the idea of waste means to a Euro-American wildlife biologist? What does it mean to the Inuit or Chipewyan hunters who take caribou? How do we reconcile radically different notions of respectfully treating caribou? These kinds of questions are painfully absent from mainstream wildlife science, where the lack of any meaningful self-reflexivity and attention to non-European worldviews has allowed biologists and managers to neglect their own politics. As Milton Freeman (1989: 106) has written:

> scientists are generally held to be purveyors of certified knowledge and are unlikely to seek to diminish their professional stature by admitting to non-scientists the tenuousness of their findings.

Caribou biologists and wildlife managers have often been engaged in the production of an imperialist knowledge – one that claims absolute authority over its subject. The subtle shift from knowledge about caribou biology to knowledge about the management of caribou is concealed in the rhetoric of their positivistic conception of scientific practice. What was conventional wisdom among wildlife biologists and what is specifically treated in this examination of Frank Miller's article, is the use of techniques that mask the contested nature of caribou biology. These include: writing histories of caribou populations so they appear to be universally accepted facts, as well as the promotion of predictions for the indigenous use of caribou without acknowledging the danger of such statistical manipulations and models. Although the caribou managers of yesterday were clearly committed to the preservation of wild caribou, their conviction in the authority of their science, the belief that their practices were objective and superior (if not the singular, rational system for knowing the world) to other ways of knowing, blinded them to the manner in which conventional assimilationist and neocolonialist ideologies of Euro-Canadians tainted their policy recommendations.

Notes

1   The importance of this website is that it is highly accessible to school, university and college students as encapsulated encyclopaedic information – which points to a larger problem with disembodied sites of official information on the Internet.

2   According to the photographer of the image in Figure 10.2, Dolores MacFarlane, the image documents the results of a 1951 autumn hunt in northern Manitoba, where sixteen men and boys harpooned caribou from canoes as they crossed a river. Ms. MacFarlane happened upon the scene when her aircraft, having experienced engine trouble, was forced to land. She camped at the spot for five days. In this particular photograph, the caribou were freshly killed – a far cry from the interpretation given in the caption of the published version. The image was used in Banfield's (1956) article as photographic evidence to illustrate his claim that indigenous hunters were wasteful, even though it showed caribou that had only just been killed. Four years later another federal biologist, J.S. Tener (1960), published a similarly accusatory paper. Ms. MacFarlane never gave permission for the photographs of caribou 'carnage' to be published in Banfield's article.

3   This is a point that is also rejected by Brightman (1993), who states that the Cree of northern Manitoba do not appear to have deliberately managed animal populations.

4   Shepherd Krech (1999) describes how Western cultures have consistently portrayed the Americas as a Garden of Eden.

5   Although their article concerns 'sedentary' (Bergerud 1988) woodland caribou – which, in comparison with migratory Barren Ground caribou, are more difficult to count due to limited vision in the boreal forests – it is at least suggestive of problems that may be inherent in a practice that is continually changing its methodology for measuring population size.

6   M.M.R. Freeman, personal communication.

11 Caribou Crisis or Administrative Crisis? Wildlife and Aboriginal Policies on the Barren Grounds of Canada, 1947–60

Peter J. Usher

The postwar years were a time of rapid change in the Canadian North. There was a growing view in government that the old fur trade economy was no longer sustainable, but what should or could be done about it was unclear. The problem seemed especially critical in the least accessible and least developed parts of the North, not least in the central Barren Grounds between the Mackenzie River and Hudson Bay. The defining event of that place and time was the so-called 'caribou crisis': the apparent confirmation by science of long-held suspicions of severe depletion of the great Barren Ground caribou herds, due to over-hunting by the Inuit and Dene who, it was supposed, were unwittingly setting themselves up for disaster.

The 'caribou crisis' was, in retrospect, constructed on relatively little hard evidence. It was sustained largely by theory, conjecture and cultural bias, and assumed such importance because it in turn gave direction to the management of both people and caribou. The problem was not merely one of caribou conservation, and it required more than conventional wildlife regulation measures for its resolution. The 'caribou crisis' provided justification not only for imposing hunting restrictions, but also led ultimately to the relocation, sedentarisation and supervision of both Inuit and Dene, who lived on or near the range of the Qamanirjuaq, Beverly and Bathurst caribou herds, and for whom these herds were not only the staple food supply but also an important source of clothing. These measures were seen by the administration as critical requirements for both the modernisation of people regarded as among the most isolated and traditional of the entire continent, and the conservation of caribou herds. The scientific management of caribou became an integral part of a broad programme of social engineering that required consensus and cooperation among various federal, provincial and territorial agencies (Figure 11.1). I review these events, especially as they unfolded on the central Barren Grounds, for the period 1947–60. I consider the nature and context of the crisis, the views of those charged with responding to it, the policies they promoted, and the outcomes.

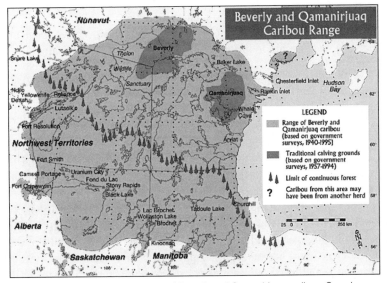

Figure 11.1 Map showing the range of Beverly and Qamanirjuaq caribou, Canada.

I have relied chiefly on the record groups of Indian Affairs (RG10) and the Northern Administration (RG22 and RG85) in the National Archives of Canada (NAC). Specific documents are thus referred to as, for example, NAC, RG22/. The notation PC refers to federal Orders in Council. The chief legislative instruments referred to are the Northwest Game Act (1917) and its successor, the Northwest Territories Game Ordinance (NWTGO) (1949).

The Caribou Crisis

The Canadian government's concern for the conservation of caribou on the Barren Grounds first arose in the 1920s, with the expansion of the fur trade and the influx of white trappers in the North. Not a half-century had passed since the demise of the plains buffalo herds, and the Dominion government was above all anxious to conserve the food supply of the Inuit and Dene so that they would continue to live on the land and not become dependent on public relief. Conservation measures adopted during the interwar period included, first, restrictions on the location of fur trade posts, the trade in caribou hides, the sale of game meat, and the entry of white trappers; second, the creation of Native game preserves from which nonaboriginal hunters were largely excluded; and third, the payment of wolf bounties as a means of predator control. No licensing requirements, quota limitations, close seasons, or gear limits were placed on aboriginal people, however. Although all persons were prohibited from killing calves, and cows with calves, this limitation was rarely if ever enforced on Inuit and Dene.

The actual need for, and appropriateness of, these caribou conservation measures were based largely on reports (often hearsay) from the 'old hands' – chiefly Royal Canadian Mounted Police (RCMP), Indian Agents, traders, missionaries, and the occasional government scientist. As would be the case for another fifty years, Inuit and Dene views on the matter were neither sought nor accounted for. But no one in authority had (or could have) actually counted the caribou, and herd distribution and migration patterns were poorly understood. There was no way of knowing whether occasional episodes of local scarcity were due to low numbers or varying migration routes. Likewise, hardly any outsider was actually in a position to observe the fall kills at the river crossings, because police patrols mostly visited the winter camps.[1] Nonetheless, allegations of 'wanton slaughter', 'excessive kills' and 'needless waste' by Inuit and Dene were recurrent themes of police reports and traders' accounts of the day.

After the Second World War, the rise of scientific wildlife management, and progress in air transportation, created new opportunities to address the issue of caribou conservation. In 1947, Canada, Manitoba and Saskatchewan entered into a cooperative three-year study of the entire Barren Ground caribou range, using aerial surveys of the entire range for the first time, under the direction of A.W.F. Banfield of the Canadian Wildlife Service (which was then responsible for wildlife research in the Northwest Territories). The study resulted in a population estimate of 668,000 animals (thought to be accurate within 20 percent), far lower than previous speculative estimates. The annual mortality rate (including human harvest, wolf kills, and other causes) was estimated at 168,000, exceeding the estimated annual birth rate by 23,000 animals. These numbers alone suggested an impending crisis, even if population trends could not yet be firmly established.

Banfield considered that the problem could be solved without resort to drastic measures, and his recommendations to the Dominion-Provincial Wildlife Conference were modest. They consisted primarily of conservation education, greater involvement of aboriginal people in wildlife administration, tighter restrictions on sales, exports, the hunting season and non-native bag limits, better kill reporting, the use of reindeer and buffalo meat instead of caribou in residential schools and hospitals, fire protection on the winter range, and experimental wolf control by poisoning. The minutes of the Advisory Board on Wildlife Protection, where caribou conservation was regularly discussed during the late 1940s and early 1950s, indicate that these recommendations fell on receptive ears. Most of these measures were put into effect in the ensuing years.

The Canadian Wildlife Service (CWS) studies, and the reports of other observers at the time, identified two causes of caribou depletion which, even if they were not necessarily the only or even the main factors, were the ones that could be most easily controlled. These were human hunting, chiefly by Inuit and Dene (although nonaboriginal hunting continued albeit on a more restricted basis since the late 1930s), and wolf predation. The first problem called for an end to both wasteful harvesting practices and the waste and misuse of meat actually taken; the second for a sustained wolf control programme. The issue of waste was at the fore throughout the 1950s, with the fall hunt (specifically, the alleged overkill beyond actual needs), and the practice of feeding caribou meat to dogs, being especially condemned.

There was more bad news to come, however. The Barren Grounds were resurveyed by air in the spring of 1955, resulting in a count of 279,000 animals, or little over 40 percent of the late 1940s count (although the Keewatin herds were reported to be stable in numbers). Two years later, the population was estimated at 200,000 on the basis of partial survey coverage (Kelsall 1968: 149–50). These results suggested that caribou were rapidly disappearing despite stricter controls, and that the crisis was more severe and urgent than previously imagined (ibid.: 283–84). Kelsall and other CWS biologists considered that radical measures were urgently required to avert a catastrophic collapse of herd populations.

The policy measures that the 'caribou crisis' inspired or accelerated must be understood in the context of another postwar crisis in the North, in the administration of Inuit and Dene. The old policy of leaving them to lead their traditional way of life, independent of government, was becoming unsustainable. Fur prices were in decline, the Hudson's Bay Company was closing posts, and independent traders were leaving the country. The cost of trade goods was rising rapidly with post-war inflation, and the need for them was increasing. Bands of Inuit began living, in miserable conditions, around military bases and weather stations for security and material goods. Tuberculosis, influenza and polio were rampant, and it has been estimated that by the mid-1950s, 10 percent of Eastern Arctic Inuit were in hospitals in southern Canada (Tester and Kulchyski 1994: 53). Reports of distress and starvation of isolated bands were reaching the southern media. In formerly inaccessible stretches of the provincial North, mines, commercial fisheries and sport fisheries were being developed. The North was changing rapidly, and both government and aboriginal people were unprepared. Beginning in the late 1940s, laissez-faire policies intended to leave Inuit and Dene as independent hunters were replaced by interventionist policies intended to bring them into the modern world, and the old colonial triumvirate of traders, missionaries and police was supplemented by government officials sent to implement these policies.

The Barren Grounds provided both opportunity and, it was thought, urgent necessity to experiment with new ideas and test capabilities in wildlife management and social engineering, on the part of a society eager to apply science, rationality and technique to peace-time reconstruction, as it had done so successfully in war.

Wildlife Management

Wildlife harvesting on the caribou range was regulated by the federal government in the NWT,[2] and provincial governments south of the 60th parallel. Canada had the power to regulate aboriginal harvesting, but provincial powers to do so were restricted by the Natural Resources Transfer Agreements of 1930.[3] Until the postwar period, however, Canada placed virtually no restrictions on aboriginal subsistence hunting. This was not so much out of regard for aboriginal and treaty rights but because a key objective of the Northwest Game Act was to ensure that Indians and Inuit could feed themselves.

The Advisory Board on Wildlife Protection (ABWLP), an interdepartmental committee established by the Dominion Government in 1916, provided advice on wildlife matters of national concern, including areas of territorial jurisdiction. The ABWLP brought together representatives of the Northern Administration Branch, the Indian Affairs Branch, the RCMP, and the Canadian Wildlife Service (among others). It was thus a body that considered the social and economic, as well as the technical and enforcement, issues associated with wildlife management.[4] These same agencies were also represented on other coordinating mechanisms for northern administration, for example, the Advisory Committee on Northern Development. Since the early 1920s, federal-provincial wildlife conferences were held annually as a means of coordinating transboundary wildlife management.

Given these administrative structures, it should not be surprising that wildlife managers became involved in social and economic policy, and that social and economic policy makers participated in formulating wildlife regulations. The minutes of the ABWLP and the federal-provincial conferences, and the internal records of the Northern Administration Branch (NAB) and the Canadian Wildlife Service (both housed in the same federal ministry during the 'caribou crisis') provide substantial insight into the views of those involved.

In this section I outline the revisions to the already existing suite of wildlife regulations. These changes, relating to access, seasons, gear, the sale of meat, sanctuaries and wolf control, were in fact not extensive. In the next section I examine in more detail the development of integrated strategies for resolving the 'caribou crisis', which were of much greater significance.

Hunting by nonaboriginals in the Northwest Territories declined mainly by attrition, as those who had qualified for the General Hunting Licence (GHL) in 1938 continued to leave the North. There were still an estimated 200 white trappers within the caribou range in 1949, and exemptions for prospectors also remained in force. Nonaboriginal NWT residents continued to be permitted to hunt up to five caribou per family for their own use. The provinces of Manitoba and Saskatchewan imposed tighter limits on northern residents and travellers in the late 1940s; these also applied to Metis, who were not recognised as having aboriginal status. Saskatchewan was prepared to eliminate sport hunting, but as a condition wanted the NWT to ban the local sale of caribou meat.

The close season, rarely if ever enforced, was removed for aboriginal people in the NWT in 1955. The use of .22 calibre firearms was prohibited for caribou hunting in 1949 in the NWT, and in 1950 in Saskatchewan, to reduce wounding losses.

Although market hunting for game meat had long been prohibited in the Provinces, it was permitted within the NWT (but not for export) on account of its isolation and lack of alternative food sources. The desirability of harmonising the treatment of this practice on either side of the 60th parallel was frequently discussed at the ABWLP and similar gatherings, but was never achieved.

Commercial sale of caribou occurred primarily at the western edge of the caribou range in the NWT. In 1947, the Fort Resolution warden's report indicated that of about 3,000 caribou killed at Fort Resolution, Rocher River and Snowdrift, about 10 percent were sold by Indians to local sawmills and the Hudson's Bay Company (HBC). Sale of caribou meat to the HBC was also reported as significant at Fort Rae. It was regarded as a way of making money when furs were scarce. Hindquarters (usually the only part sold, with the rest being consumed domestically or fed to dogs) sold for $1.25. Caribou was also sold in Yellowknife, where gold-mining operations had recently been revived.[5] In that same year, new regulations prohibited the sale of caribou meat 'at hotels, restaurants, or other establishments where a charge is made for meals,'[6] but trade in meat among individuals remained legal.

The system of Native Game Preserves, which had been expanded from the 1920s to about 1945, fell into disuse. The Arctic Islands Game Preserve (which included the northeastern mainland) was eliminated in 1966 (Hunt 1976). The repeated requests by Saskatchewan Dene chiefs for a 'reserve' (by which they meant an exclusive Dene hunting preserve between approximately latitudes 58 and 64 degrees North) were, despite sympathetic responses from field officials, ultimately ignored. During the 'caribou crisis' itself, however, no changes occurred. Expanding the preserve network would have had little practical effect as nonaboriginal trappers were abandoning the Barren Grounds at the time, the sport-hunting industry was not yet developed, and, with improved air transport, mineral exploration parties were better able to bring their own provisions.

The wolf bounty was abandoned in the early 1950s as ineffective. Manitoba experimented with poison baits in 1949, and, after some refinement, a wolf-poisoning programme was instituted in direct response to the caribou crisis by all wildlife agencies on the caribou range. It continued into the early 1960s, when it was considered no longer necessary (Kelsall 1968: 254–56).

Integrated Policy Initiatives

The four most important integrated policy measures in response to the 'caribou crisis' were: limiting institutional use of caribou; reducing waste of caribou meat; education and enforcement; and perhaps the most far-reaching, restricting the aboriginal harvest. Each of these had implications much greater than wildlife management itself and required consensus and cooperation of several agencies and jurisdictions. Implementing these measures would also require more aggressive and intrusive social engineering among Inuit and Dene. I explore the views expressed by administrators, wildlife scientists and enforcement officials as these initiatives were developed.

Institutional Use

The issue of serving caribou to aboriginal people in residential schools and hospitals became controversial in light of the 'caribou crisis'. The amount of meat involved was significant, in view of the rising number of aboriginal persons in those institutions the late 1940s. For example, the Roman Catholic Mission at Fort Smith was reported to have bought 400 hindquarters in 1949.

The NAB proposed to restrict the amount of meat supplied to institutions by imposing quotas and permits for hospital use, and eliminating the supply to schools. In the Mackenzie District, increased quantities of buffalo meat from Wood Buffalo Park would be supplied as a replacement.[7] The ABWLP recommended a transitional allowance until increased supplies of reindeer and buffalo meat, and improved cold storage facilities, became available. In response, the RC Mission disputed both the proposed allocation and the reasons for it, and in some cases ignored the regulations. Although these infractions were noted, charges were never laid.

The Bishop told the government that

> Even when the people are sick, and find other food distasteful, they will still eat the wild meat. They know there is an abundance of it, from time immemorial, and there will be until Divine Providence provides something else for this North country, as it has been provided across the prairies, where wild game was formerly without number. Actually the quantity of caribou meat we require for our institutions is less than a drop of water to a lake, when compared with the numbers of caribou now roaming through the North. To refuse those sick people that meat is inhuman.

He added that buffalo meat was disliked by patients, and it would be unacceptable in the south to deny patients the food they craved and substitute what they disliked. If patients were at home, no one would deny them caribou meat.[8]

Commenting on the Bishop's view of the benevolence of Divine Providence:

> Mr. Wright mentioned to His Excellency at Fort Smith that no comparison could be made as between the Northwest Territories and the prairies inasmuch as the former was not suited to agriculture and that if the meat supply disappeared the natives would be in a bad way. The Bishop stressed that he was quite happy to leave it to a benevolent Providence to work out a solution. This is brought to your attention as a clue to the attitude of the Bishop towards attempts of the department to follow the advice of wildlife investigators.[9]

Evidently the Catholic Mission did not share the administration's enthusiasm for rational, scientific management of either caribou or aboriginal people. The Mission came to be viewed by the NAB as uncooperative with respect to the game laws, and encouraging 'backwardness' in Indians.

Government officials placed little importance on aboriginal food preferences, against the need for conservation. Indian Health Service officials, for their part, stated that preference for game meat was a matter of taste, that caribou was not necessary to the welfare of inmates if a balanced diet were provided, and that patients taken to southern institutions for treatment had quickly adapted themselves to the general diet. Serving game meat in mission hospitals was therefore a luxury, and on the basis of scientific advice the regulations must be enforced.

When a Fort Resolution hospital patient wrote to the NAB in Ottawa, asking why they were forbidden to eat caribou meat in the sanatorium and noting that 'We are tired of buffalo meat and can[ned] stuff, as we didn't live on these foods before, so it has become quite a change for us all, leaving a poor appetite which is bad for tuberculosis people', he was told that 'it is not any real hardship to have to do without caribou meat and particularly when you know that by doing so you are going to help your children and their children.'[10]

In the Keewatin District of the NWT, restrictions on institutions were even more severe, despite the fact that the government was even less capable of supplying alternative sources of meat there. In 1949, the ABWLP recommended against an application by the Catholic Mission to provide caribou at the hospital and industrial home at Chesterfield Inlet.[11] Under the game regulations, the mission itself could not take more than five caribou in each settlement in which it was established.[12] A Commissioner's permit further authorised the use of twenty-five caribou in 1954–55 for the residential school, but not for use in the hospital or the industrial home,[13] with the advice that 'The Missions are naturally considered a most important influence in the communities they serve and may be counted on, we hope, to set a good example in observance of the law.'[14] The NAB took the view that it could not make an exception for Chesterfield without doing the same for the Mackenzie River missions.

The Catholic Mission then took the matter up with the Minister, requesting a change in the Game Ordinance. The letter noted that healthy Eskimos were allowed to hunt for food and clothing all year, and that the only object of the present Act to was to safeguard caribou herds so Eskimos could maintain their traditional economy. When

Eskimos were sick and sent to hospital, they were penalised because the hospital was managed by white people and they were thus deprived of the food most familiar and wholesome to them. Chesterfield had an average of fifty-five patients throughout the year, and 'supplying them with food other than caribou meat is very expensive and above all it does not bring them the physical and psychological welfare to which, as sick persons, they are entitled.'[15]

No change was forthcoming, however. The NAB took the firm view that unless there was clear evidence that the caribou population was increasing, there was no basis for relaxing the present policy, even though key wildlife scientists in the field recommended leniency.[16]

Waste

Up to the end of the 1940s, no solid evidence had ever been amassed to the effect that the Barren Ground caribou herds were being depleted by overhunting or waste by aboriginal hunters. There persisted, nonetheless, a widespread view that 'waste', in the form of 'needless', 'wholesale' or 'ruthless' slaughter, was a troublesome and threatening phenomenon. Opinions varied on who was most responsible for it: Indians, Inuit or white trappers, but the consensus usually went in the direction of the Denesuline (northern Chipewyan), especially the Maurice and Barren Lands Bands. The matter was commonly raised at meetings of the ABWLP, usually on the basis of police reports.

There cannot be much doubt that kill levels were substantial. Banfield (1954) estimated 125 animals per aboriginal hunter per year, of which perhaps half were for dog feed (per-hunter kill levels may have increased in the early twentieth century because, as people were drawn into fur trapping, they tended to use more dogs). Rough estimates from the 1920s and 1930s, based on anecdotal accounts, were even higher, but Canadian Wildlife Service estimates of kill levels for 1950 were substantially down from a decade before (Anon. 1982).

Whether these kill levels, whatever they actually were between 1917 and 1947, constituted waste in the sense of an unsustainable demand on the herds, has not been demonstrated. We may never have a conclusive answer to this question, or to whether the herds actually did decline during that period. There is a lot of evidence, however, that non-aboriginal observers were culturally predisposed to see waste in situations that Dene and Inuit were not.

Waste of harvested meat, as opposed to excessive harvesting, was not widely regarded as a problem until about 1950. The only references in the files prior to that time are with respect to dogs, and the possibility that their numbers were excessive and therefore making too great a demand on the resource.[17]

When revisions to the Northwest Game Act were being considered by the ABWLP in April 1947, it was considered permissible for natives to feed caribou to dogs where necessary,[18] which at that time was the rule. The Act only prohibited destruction or spoilage of game meat suitable for human consumption.[19] Banfield, in his first progress report to the Federal-Provincial Wildlife Conference of 1949,[20] included both excessive harvest and unutilised harvest under the subject of waste. He considered the main sources of waste to be wounding with .22s, waste of meat from the summer hunt for hides, killing more animals than needed, loss of meat to predators from unprotected caches, and use of carcasses for bait. He did not mention feeding meat to dogs.

The administration soon took up Banfield's views and added to them. In 1950, its publicly stated view was that waste was due to killing more than could be utilised, wounding with .22s and nonretrieval, too many dogs, and excessive slaughter.[21] Reports coming in from the field were mixed, however. Reports of substantial waste of caribou meat at Chipewyan camps were cited at the ABWLP meeting of 17 August 1950, but a 1951 warden's patrol to Rocher River found no evidence of waste. According to the Indian Agent at Fort Resolution, local Indians sold the hindquarters, used the forequarters for themselves, and fed the rest to their dogs.[22] At its August 1950 meeting, the ABWLP endorsed a recommendation to prohibit feeding of any part of a caribou to dogs at settlements where other dogfeed was available, although members recognised it would be difficult to enforce.[23]

At the same time, another strategy emerged. This was the provision of cold storage facilities in the Mackenzie River communities to enable the import and storage of fresh meat and relieve hunting pressure on big game.[24] In that year, the Indian Affairs Branch sent trial refrigeration units to Fort Chipewyan, Fort Resolution and Yellowknife, to preserve native foods for redistribution in times of scarcity, with the objective of installing several more if these proved successful.[25] However, this strategy was not regarded as feasible in the Keewatin. The Chief of the Forests and Game Section recognised that storage and caching was a problem in Eskimo areas, but while the Branch could encourage better caching and discourage carelessness, 'we cannot at present insist on a nomadic people carrying out careful storage.'[26]

Following Kelsall's 1955 report indicating a continuing decline, dealing with the waste problem became a higher priority and was seen to call for more aggressive measures. Sivertz, outlining the options for the Commissioner, noted that there were already provisions in the NWTGO prohibiting waste, but that these were difficult to enforce. He noted the Indian Affairs Branch policy of supplying refrigeration units to Indian communities (within the caribou range, Stony Rapids had already been added, and units were under construction at Fond du Lac and Snowdrift), and noted that there was nothing like this in Eskimo territory, except at the Catholic Mission at Chesterfield. He suggested constructing units at several places in the Keewatin, even while noting that underground cellars in permafrost would also suffice and that Inuit could be employed in their construction.[27] The NAB also proposed an increased slaughter of buffalo from Wood Buffalo National Park, and the distribution of low-cost buffalo meat to Mackenzie Valley settlements.

When the Technical and Administrative Committees on caribou were established in the fall of 1955, the top priority was to eliminate waste. All present at the meeting agreed that this was still a problem with both Indians and Eskimos, due to improper caching, feeding caribou to dogs when alternatives were available, and 'wanton killing with carcasses left to rot.' The last was considered most common at Duck Lake and Brochet in Manitoba, and Indian Affairs proposed to place a man in this area to attempt to curb it.[28] In Saskatchewan, local game officers organised fall fishing parties to promote Chipewyan to feed fish instead of meat to their dogs.[29]

A paper prepared for the NWT Council in 1959[30] outlined in some detail the measures undertaken by the federal government to reduce the demand for caribou meat, consisting chiefly of importing game meat, promoting fishing and marine mammal harvesting, and expanding refrigeration capacity. The Indian Affairs Branch was distributing buffalo meat as a relief measure in the NWT and in northern Alberta and Saskatchewan. Of a planned distribution of 85,000 lbs. in 1959–60, about one-quarter was destined for Dene communities at the western end of the caribou range in the NWT and Saskatchewan. About an equal amount of elk meat (culled from Elk Island National Park in Alberta) was to be shipped to Churchill, at the eastern end of the range. The Indian Affairs Branch issued fish nets to Indians, Northern Affairs organised fishing, sealing, and whaling projects in the Keewatin, and the Department of Fisheries conducted stock surveys of inland lakes in the Keewatin. The Indian Affairs programmes continued at least into the mid-1960s.[31]

By the early 1960s, there was a growing perception that the waste problem was declining, even in northern Manitoba, although it was still reported from time to time, and occasionally investigated.[32]

Education and Enforcement

Education had been seen as an essential element of wildlife conservation in the far north since the 1920s. Whatever form education campaigns took in those early years – propaganda, exhortation or threats – and whatever success they may have had, there was little alternative. There was virtually no effective enforcement capacity in any jurisdiction. Enforcement was mostly delegated to the RCMP, which maintained detachments (in some cases discontinuously) at Rae, Yellowknife, Reliance, Stony Rapids, Churchill, Eskimo Point, and Baker Lake. The NWT did not establish a separate warden service until the late 1940s. There was a Saskatchewan Provincial Police detachment at Fond du Lac briefly in the 1920s, and from about 1950 onwards, Manitoba stationed a Provincial Conservation officer at Brochet. All of these points are at the edge of the range; both Dene and Inuit spent most of their time on the land, in the heart of the range, far away from these places. Patrols were made at most annually or semiannually, usually by dog sled, even long past the days when white trappers started going into the country by air.

Throughout the early 1950s, there was substantial debate within and between the Northern Administration Branch and the Canadian Wildlife Service about the balance between education and enforcement. On balance it seems that the CWS, and wildlife biologists generally, leaned toward enforcement, while the administration leaned toward education.

At the ABWLP meeting of November 1947, it was observed that the lack of a field force capable of enforcement had led to a lack of compliance. There was a need to train both staff and Natives, and it was hoped that in three to four years, Natives would learn to observe the game laws. Enforcement was progressively stepped up in the Mackenzie District, for example with emphasis on seasons in 1946,[33] and on the restrictions on the sale of meat in 1950.[34]

The problem was again discussed in February 1950. The RCMP and the administration favoured leniency, especially in the more isolated areas where people were almost totally reliant on game. There was general agreement that Natives should not be forced to comply until there was more education, an improvement in economic conditions, and an effective substitute for caribou skins.[35] At the ABWLP meeting the next month, Banfield noted that:

> It is necessary, therefore, to have suitable regulations for the protection of wildlife but in the administration of the regulations a liberal interpre-

tation had customarily been placed upon the provisions which affected the well-being of the native population. The native was privileged to kill game to prevent starvation and it was often necessary for him to kill game for food purposes out of season. There is no case on record of a native being punished for taking game animals contrary to the regulations when it was proved that an emergent [sic] situation had developed and it was necessary for him to take such game.[36]

At its next meeting, the Board endorsed the recommendation of field officers that there should not be automatic cancellation of a GHL if the holder were convicted under the Game Ordinance. This penalty was regarded as too severe as it could deprive persons of their livelihood, and it was felt that cancellation should be at the discretion of the Magistrate or Justice of the Peace at trial.[37]

In early 1951, wardens posted signs in the Fort Smith district warning that violations of the Ordinance with respect to feeding caribou to dogs would be prosecuted,[38] and in November, consideration was given to laying charges in a case of feeding caribou to hospital patients, but the matter was regarded as particularly sensitive and did not proceed.

Although education was regarded as the necessary and primary tool for conservation, concerns were expressed over its effectiveness. For example, at the officials' meeting in February 1950, the value of poster campaigns was questioned. The newly reorganised Northern Administration and Lands Branch (NALB) had just published *The Book of Wisdom* in Inuktitut, but did not have the resources to employ supervisors in Native hunting camps, as had been suggested by Banfield.[39]

The Commissioner of the NWT (and Deputy Minister of Mines and Resources) noted, in a draft memorandum entitled *Education Can Help Save the Caribou*, that:

it is obvious that conservation cannot be taught by any form of coercion or regulation. The necessary restraint can only be secured, and the co-operation of the hunters enlisted, by explanation and persuasion.

Government agents, missionaries, teachers, traders and others, who live among the natives and have their welfare at heart, have a special responsibility in regard to educating the hunters in conservation. Such persons are asked to make clear by patient and continuous education how necessary it is for the hunters to kill caribou in moderation .... Those who teach in the schools and missions should make conservation of wildlife a part of the daily educational program, bearing in mind that the children of today are the hunters of tomorrow.

While some hunters are improvident, others are practical conservationists. Most of them will co-operate gladly if they are convinced of the facts and the need.[40]

The emphasis, he stated, should be on teaching people not to kill more than they need, not to feed meat to dogs (and indeed to get rid of excess dogs), to cache meat carefully, and to use other foods like fish when available. The published version was sent to persons residing within the range of the caribou and whose assistance could be beneficial.[41] These efforts were supplemented, in 1951, by the distribution of booklets and circulars, a motion picture, and a film strip.

However, as the caribou situation appeared more and more critical, some questioned the appropriateness of the education strategy, and even its priority over enforcement. Chief among these seems to have been Kelsall, who was then in charge of the CWS caribou research programme in the NWT. In a 1954 memorandum entitled *Education*, he commented on educational efforts to date, and asserted that, after thirty years, they were not working. He believed that the Department's films and film strips had had considerable circulation in the more 'civilised' areas of the North (by which he meant the Mackenzie River district), thanks to the initiative of individual wardens, missionaries and teachers. However, he felt that many of the Natives who saw them did not follow or retain the basic ideas after only one or two showings. While police and wardens had been instructed to convey conservation information to Natives, he thought this was being done in a haphazard and frequently ineffective way: 'Many of the persons involved are not equipped by nature or training to be efficient in such work and many refrain, sometimes wisely, from taking any action whatever.' Pamphlets and talks were not enough, Kelsall argued, and he urged that conservation should be taught in the schools. The curriculum should include training in efficient hunting and trapping, care and handling of fur and meat, and care and use of firearms. He drew attention to some of Banfield's unimplemented recommendations of 1950, including the hiring of special personnel to instruct Natives in their camps, and the employment of young Natives of superior ability as assistant game officers.[42]

In view of the situation, Kelsall also called for improved enforcement. He claimed that infractions relating to seasons, permitting, gear, waste, and feeding caribou to dogs, were frequent, but also frequently overlooked. He attributed this partly to game officers overlooking infractions because 'in the Northwest Territories, and especially in Eskimo country, many persons are forced to break the game regulations in order to maintain themselves and their families'. He noted that fifty-four charges had been laid under the Act from 1948 to 1953, mostly in the southern Mackenzie, with eleven in the Fort Smith district and eight in the Fort Resolution district (there were by then still no wardens east of the Slave River, and no enforcement or charges by the warden service on the caribou range itself).[43] Most of these charges, he observed, were brought by the RCMP, and many wardens had never laid a charge. This, he said, was due to a lack of training in proper procedure, and the lack of adequate transport, as in some cases wardens did not even have their own dogteams (winter patrols by dogteam would continue to be routine for the RCMP for nearly another decade). Kelsall advocated enhanced enforcement with respect to waste and abandonment, feeding caribou to dogs, and the use of caribou as bait, although he acknowledged that the last was now rare compared to the peak of the practice in the 1930s by white trappers.[44]

The NALB, commenting on Kelsall's report, regarded draconian enforcement as counterproductive. The Chief of Forestry and Game stated that '[t]o arbitrarily exact enforcement of what appears, from the viewpoint of relatively primitive people, to be very strict legislation, would lead to deep resentment and non-cooperation and would probably damage the cause of conservation for years to come.' Management, he suggested, required enlisting the support, cooperation and understanding of trappers, which could only be done by giving them greater responsibility in management in their areas, and a voice in the development of new game laws. This had been tried and proven in some provinces, whereas in the NWT, 'when proposals for legislation were being considered, the trappers did not have an opportunity to express their views and felt they did not share in it'. He went on to assert that 'natives do not consider a jail term for an infraction of the Game Ordinance any hardship or disgrace. In jail they are very well clothed, well fed and well looked after according to their standards. Their work is light and their families are generally maintained through the issue of relief rations. A jail term under such circumstances is a picnic.'[45]

Burton's superior endorsed his preference for education over enforcement, but also noted the difficulties:

we must direct the minds of these people out of the deep channels in which they have been travelling for centuries by bringing to them entirely different concepts. Once this has been done, then I think we can start to instil into their minds some of the ideas of civilised man, which society, because of densities in population, have had to adopt for their own preservation. One of these is the careful managment [sic] and conservation of wildlife resources. We have numerous reports to show that every attempt to educate these people in game conservation has failed and although no one has given the reason for that failure, I think that we can infer that the reason has been that conservation education has been introduced too early in the overall education of these primitive people.[46]

Within the year, the warden service had been relieved of its enforcement responsibilities, with these being returned to the RCMP by mutual agreement. This left the warden service with administrative, advisory and education functions which, it was considered, it would be better able to carry out without the encumbrance of being perceived as an enforcement agency. Nonetheless, it would supply the RCMP with evidence required to secure convictions. The relationship between the warden service and the RCMP in the Mackenzie District would thus become very much like that in the western provinces.[47]

In the NWT, education campaigns consisting of talking to hunters and trappers while on patrol, speaking at meetings and at schools, and booklets in Inuktitut (although apparently not in Chipewyan), and sometimes coupled with threats of enforcement, continued in the late 1950s and into the 1960s.[48] These campaigns involved, as they had in the 1920s, the RCMP, the northern administration, the Indian agencies, and the more recently added game wardens. There were continuing calls for immediate clamp-downs. Kelsall, for example, asserted after the 1957 survey that without such enforcement, there would be no caribou left to resurvey.

Manitoba indicated a desire to implement an education campaign in 1949, coupled with enforcement, particularly on non-Natives in the North.[49] Provincial Conservation Officers seem to have favoured enforcement, but lacked the resources to do so. In the early 1950s, the Conservation Officer at Brochet submitted several reports on the problem of waste. During a patrol to the Seal River, he encountered a case of feeding caribou meat to dogs at a Denesuline camp. There had been, he said, no effort by these trappers to put up fish, and the nets issued to them had not been used. Having issued warnings on the matter for years, he said, he claimed now to have enough evidence to convict them. He therefore suggested getting an arresting party to come in by plane and take them out to the Pas. There, he said, they wouldn't be able to raise money to pay the fine, and would have to serve a sentence. If tried in Brochet, 'the gang just chip in and pay their fines and they think its a great joke. ... The hardest punishment that could be dealt to a Chip would be to lock him up where he couldn't talk to one of his kind for sixty days or so; he would be ready to work when he got back.'[50]

The same officer (who spoke little or no Chipewyan) reported in a similar vein on a meeting with the Nueltin Lake trappers, while on patrol in that area in March 1953: 'I have just about run out of threats and warnings, but I have them all leaving for Nueltin Lake in the morning. They are going to leave their families here and make a trip up there to pick up their mink traps. I feel helpless here dealing with these people, if they called my bluff it would put me in a tough position. If there aren't steps taken to clear up this type of situation very shortly, there will be no use in keeping a field man here.' His dislike of the Denesuline of the area is evident from his patrol report of November 1953: 'It is quite evident that the Nueltin and Fort Hall Chips, don't intend to improve their conditions, they still will not put up fish for dog feed or build proper meat caches, they still go in for the big slaughter of caribou by spear, and its certain that they are too lazy to dry this meat and look after it, so there is still a lot of waste at these two camps.' It would appear the feeling was reciprocated, as he reported that the people had moved to Duck Lake because he was too tough on them. The leader of another nearby camp, he recorded, had 'sent word down to me this fall, that it would be no use in me travelling in his area this fall, as they wouldnt give me any dog feed. Camped here for the night these guys all sat in their shacks and wouldnt even come out to talk to me none of them would sell me any dog feed, so I had to cook rice for my dogs. I couldnt see any sign of fish put up at this camp and very little sign of meat.'[51]

Petch (1994: 30) ascribes the use of 'sensationalised photos' of caribou kills in mid-1950s to Manitoba Conservation Officers, and regards them as the promoters of restrictions and sanctions on the Dene harvest, to which the scientists and government officials willingly responded (see also Figure 10.2 in Campbell's chapter). These measures are said to have included asking local Indian Affairs agents to issue smaller amounts of ammunition to Dene hunters, and imposing penalties for excessive or wasteful caribou utilisation.

Saskatchewan seems to have promoted conservation education through its Fur Conservation Program, begun around the same time as the reorganisation of trapping by fur blocks beginning in 1948. Some confiscation of .22s occurred at Uranium City, but apparently not from Indians, as there was no legal authority to apply this measure to them (Cranstonsmith 1995: 123). At a 1953 meeting of CWS, Indian Affairs, and provincial wildlife agencies to consider the caribou issue, a need was recognised to 'develop a sound workable educational conservation program and put it into operation'.[52]

While there was some sentiment for enforcement, and sometimes very tough enforcement, it was restricted largely to biologists and field officers, usually Conservation Officers but sometimes Indian Agents. There is not enough evidence to say whether those were the *prevailing* views of such persons, but they existed and would certainly have been communicated to aboriginal people, particularly Denesuline. However, such agitation for enforcement measures generally went unsupported. There was, on the other hand, tighter regulation and perhaps stricter enforcement of commercial harvesting regulations, especially for fur trapping, which was seen as a privilege rather than a right under treaty. Again this would have applied to Denesuline more than Inuit. There may have been some fur seizures but further research is required to verify this. There is a substantial record of objection and even resistance to regulations and enforcement by Denesuline, almost continuously from the signing of the treaties.

In any event the capacity for enforcement was virtually nonexistent until the late 1940s, and even then, only in the western part of the range, and it was always under-resourced. Aircraft patrols were virtually unheard of. Even so, neither the CWS nor the NALB had sufficient personnel for either education or enforcement. There was a continuing reliance on the RCMP, as well as on the agencies on the ground, such as the missions and, beginning in the late 1940s in the west, the schools. The HBC seems not to have been regarded as a useful ally in this regard, as in many quarters there lingered a hostility towards their profiteering on fur and wildlife, and, as noted above, there were also doubts about the Catholic Mission.

Another difficulty with conservation education, however well-meaning, was that few nonaboriginals in authority had any idea of how to communicate effectively with Inuit or Dene. Clancy suggests, probably accurately, that 'conservation education may have meant little more than upbraiding the natives for careless hunting and waste of game' (Clancy 1987: 12). The method was nonetheless effective insofar as threat of sanction often promoted compliance, obviating the need for the more problematic and expensive judicial route of charges and convictions. RCMP officers often threatened sanctions (with or without any legal basis for enforcement) as a means of getting Inuit in isolated communities to comply with the game laws, send their children to school, or otherwise do what the police wanted them to do.[53] Thus while the more extreme views noted above seldom prevailed at the policy level, they were often the ones that Inuit and Dene actually heard (to the extent that they understood them).

Limiting the Aboriginal Harvest

Perhaps the most extreme measure considered was to limit the aboriginal harvest itself. Sentiment for doing so had existed since the 1920s, especially with respect to Indian treaty hunting rights. The problematic nature of these rights in the eyes of provincial wildlife managers was raised as early as 1922, at the first Federal-Provincial Wildlife Conference. One of its resolutions, noting that wildlife was an important national asset and that 'Indians and others if not restricted will eventually deplete the supply', and further noting the 'liberal' provisions for Indians in northern Canada regarding the taking of game for food at all seasons, called upon the Department of Indian Affairs to 'continue to point out to all such Indians that in their own interest and in the interest of the country that the Provincial and Federal Game Laws be observed'; and further that 'when in the opinion of game officials of any section of Canada, it is considered necessary to further restrict the killing of game due to a decrease in the supply' that DIA officials cooperate with them in their efforts to conserve.[54]

Wildlife managers in the Prairie provinces believed that the Natural Resources Transfer Agreements of 1930 had only exacerbated the problem. A resolution of the 1932 Provincial-Dominion Wildlife Conference noted that whereas 'the Natural Resources Agreement between the Dominion and the Prairie Provinces provides opportunity and excuse for excessive killing of game and other wild life by Indians in those Provinces', it was resolved that while sympathetic to the actual needs of Indians for wildlife, the conference was

> of the opinion that the existing provisions of the [NRTA], if literally carried out, will cause serious depletion and possibly practical destruction of game in the provinces concerned, resulting in great distress for the Indians, and therefore urges that any interpretation placed upon either any Indian Treaties or the Natural Resources. Agreement should be based upon the necessity of preventing widespread extirpation of wild life and that in the best interests of game and of the Indians this interpretation should be consistent with the generally recognised reasonable principles of conservation and perpetuation of valuable game and other wild life resources.[55]

The conviction that Indian hunting rights should take second place to scientific wildlife management, and that this was for the Indians' own good, would be repeated in the decades to follow. Following Banfield's studies, there were frequent assertions to the effect that nineteenth century treaty guarantees were outmoded in the light of current conditions.

Quotas, which to date had not been imposed on Inuit or Dene anywhere in the caribou range, were seriously considered in the NWT in the mid-1950s, although ultimately not adopted. Later, the idea resurfaced in the form of proposals to restrict access to caribou by Inuit having waged employment and therefore better access to alternative food sources.[56] Although again not adopted, the idea of limiting GHLs to persons without permanent employment continued to be discussed by the NWT Game Branch into the mid-1960s.

In 1957, the CWS proposed that caribou be declared in danger of extinction, and therefore subject to special regulations. The NWT Council did so, but a subsequent NWT judicial decision questioned the applicability of the NWT Game Ordinance to Inuit. The federal Northwest Territories Act was amended to ensure that it did, but at the same time barred restrictions or prohibitions on 'Indians or Eskimos ... hunting game for food on unoccupied crown lands, other than game declared by the Governor-in-Council to be in danger of extinction' (Clancy 1987: 22). A federal Order-in-Council declaring caribou, muskox and polar bear in danger of extinction was passed in 1960.[57]

As this measure could not apply to Indians in the provinces, it was further proposed that the Natural Resources Transfer Agreements (NRTA) be amended to enable similar restrictions there with respect to endangered wildlife. Negotiations to this end proceeded among federal and provincial ministries for two years (1960–62), with draft legislation proposed to the federal cabinet.[58] Ultimately the proposed amendment did not proceed, chiefly because of the Sikyea hunting rights case in the NWT, and interventions by the Federation of Saskatchewan Indians (Cranstonsmith 1995: 127). Nonetheless, this policy continued to be promoted by the CWS, and particularly Kelsall (whose preferred option was a complete cessation of all caribou hunting for several years), into the mid-1960s (Kelsall 1968: 285–86).

Policies Implemented

Most of the integrated policy initiatives were not implemented in the way that they were proposed. The most effective was the limitation on institutional use, this being the most amenable to government control. Inuit and Dene harvesting, and the use of caribou for dogfeed, seems to have declined substantially from the late 1940s to the mid-1960s, but not because of outright harvest limitations or the effectiveness of the regulations as such. It would also appear that conservation education programmes, such as they were, were largely unsuccessful. On the other hand, the 'caribou crisis' certainly contributed to the relocation and sedentarisation of Inuit and Dene, although this was more consequence than objective of the remedies originally proposed.

Relocation of aboriginal populations on the range (or elsewhere) was not new. The Hudson's Bay Company established the Caribou Post in 1930 to keep the Sayeze Dene inland and away from civilisation. Those people moved to Little Duck Lake in 1941 when a US Air Force weather station was established, and the Ahiarmiut moved to Ennadai Lake in the mid-1940s when a Canadian military radio station was constructed there. The NAB promoted a relocation of the Ihalmiut to Nueltin Lake in 1947 to participate in a commercial fishery, and provided emergency rations there, but most returned to Ennadai.

As famine reports increased, relocation became the preferred government response as the most reliable method of supplying food. The Ennadai Lake people were moved to Henik Lake in 1956, and the next year to Whale Cove and Rankin Inlet on the coast.[59] This last move was justified in part as a means of allowing caribou to recover. By the mid-1960s, the relocation of a substantial portion of the Inuit population out of the Keewatin territory altogether was given serious consideration within the Department of Indian Affairs and Northern Development, although never implemented.

The Sayeze Dene were also relocated during this period. The Duck Lake post was closed in 1956, and the Dene were evacuated to Churchill suddenly that summer, at least in part, some have suggested, because the Manitoba Game Branch wanted to ensure that the fall caribou kill would not take place. They remained in Churchill under appalling conditions for fifteen years before moving back to Tadoule Lake. The Barren Lands Band was encouraged to settle at Brochet in the mid-1960s, after which the people no longer spent the summer on the Barren Grounds as they formerly had done. In Saskatchewan, trapline registration in the late 1940s facilitated sedentarisation at Fond du Lac and Wollaston.

Thus, on the eastern part of the range, the key Inuit and Dene hunters were moved to the edges of it and sedentarised. No longer did they organise their movements around the caribou, especially at migration time when large numbers of animals could be killed. The Barren Lands Band and the Saskatchewan Dene bands now spent the summers south of the 60th parallel, in the provinces, and went hunting and trapping only after the snow came. The Indian Agencies supplied the people with fish nets, encouraged them to engage in the commercial fisheries on the large lakes in summer (and in guiding in the newly developed sport fisheries), and organised summer fisheries for dogfeed. They also installed walk-in freezers on the reserves.

In the Keewatin, the Ahiarmiut were removed to the coast, and encouraged to fish and hunt marine mammals. Whale Cove and Rankin Inlet became centres of 'modernisation', where services could be provided, a new economy built on commercial resource harvesting, education, training, and industrial employment. Rehabilitation became a social as well as a medical concept. New housing meant new programmes to train people how to live in them, which meant that they needed to learn modern concepts of hygiene. Country food was only desirable to the extent that it could not be replaced by southern foods. These programmes were implemented with particular enthusiasm following the creation of the Department of Northern Affairs and National Resources in 1953 (which housed both the NAB and the CWS). Gordon Robertson, the new Deputy Minister, was from an early date convinced that the old economy was dead, and that the future lay in training and industrial employment (Robertson 1961).[60]

These changes were by no means universally supported by the 'old hands', who often expressed concerns about the development of a welfare mentality and 'improvidence', and the discouragement of Inuit and Dene from 'hustling for themselves'. Nor did the 'old hands' always welcome the newly hired Northern Service Officers or facilitate their work. In this environment, it is not surprising that wildlife scientists in the field had, and expressed, their own views on these issues.

There was little disagreement, however, on the need for 'supervision'. The notion that aboriginal people needed supervising was certainly not new, and especially not with respect to wildlife harvesting. Indeed Banfield's first recommendation, that field officers be employed in native camps to instruct them on conservation and resource use, was all about supervision. But this proved impossible to implement on the ground, especially in the eastern part of the range. There were not enough staff, it was impractical to station monitoring or enforcement personnel in the seasonal camps, and there were not enough aircraft support for patrols or enforcement (contrary to the situation in Alaska during the 1950s, as described by Burch 1995). At the western end of the range, in the Mackenzie District, there was a trend towards 'supervised hunts' using aircraft, made both possible and necessary due to seasonal sedentarisation. These also had the effect of limiting the kill to the number of carcasses that could feasibly be brought back to the communities by air (Kelsall 1968: 203).

Relocation and sedentarisation accomplished the same ends as supervision in the field, and provided convenience in delivering services and administration. While the 'caribou crisis' was not the sole and perhaps not even the primary cause of relocations, even in the most dramatic cases such as the Ahiarmiut and the Sayeze Dene, it meshed conveniently with the administrative crisis and became a point of mutual support between the CWS and the NAB. It seems clear that the CWS readily came to support relocation as a means of reducing the harvest, perhaps especially in view of its inability to impose harvest quotas at will.

Although Cranstonsmith argues that the wildlife scientists' recommendations went beyond the realm of science (1995: 96), this was not unusual given the administrative structure of the time, and the fact that wildlife management was then, as it is now, as much a matter of managing people as managing animals. It is perhaps more remarkable that CWS policy prescriptions went so far beyond the normal wildlife management tool kit, but, by the same token, the Northern Administration Branch had very significant input to wildlife management policy.

Supervision in town remedied the inability to patrol and supervise in the field, and the CWS fully supported the removal of Inuit and Dene from the range. Sedentarisation facilitated food replacement, and by effectively restricting the summer and fall caribou hunts on the range, reduced the supply of material for winter clothing. This in turn reduced the feasibility of extensive winter travel for trapping and hunting, and hence the need for large dog teams. Caribou harvests by both Inuit and Dene declined substantially once people became confined to the communities.

Thus relocation and sedentarisation, although not fully developed as a policy at the outset, proved the most effective solution to the 'caribou crisis'. And so the crisis faded (although did not disappear entirely) from the administration's view. The cost to Inuit and Dene who inhabited the caribou range was high, however, as their way of life was brought to an end in haste by highly intrusive administrative actions they could neither understand nor influence.

Epilogue

The influence of scientific wildlife management, and of wildlife management agencies, on the social and economic life of Dene and Inuit, and on their legal rights, declined rapidly in the 1960s. The consensus on the causes of and cure for the 'caribou crisis', which was given considerable popular exposure (Banfield 1956, 1961b), began to unravel, and dissenting voices were heard increasingly both within and outside of wildlife management circles (Cranstonsmith 1995: 128–36). In retrospect it may be impossible to determine whether the threat of caribou depletion was real. It may have been a product of the survey methods and the knowledge of caribou biology as it existed at the time.[61] Low numbers, if real, may have been a low point in what is increasingly recognised by caribou biologists as a long-term cyclic variation in caribou populations. By today's standards, the scientific evidence for the crisis was flimsy indeed.

Caribou research methods continued to develop. Direct observation and census by aerial survey were supplemented by handling and tagging of the animals themselves to confirm movements and herd delineation, beginning in 1959 in Manitoba and extended to the NWT in the early 1960s. Aerial census methods themselves became more systematic, issues of sampling and observer error were addressed, and in the 1980s it was discovered that photo counts produced higher population estimates, all of which led to reassessment of the earlier counts.

Aboriginal people themselves began to challenge the analysis and prescriptions of scientific wildlife management. Provincial and national Indian organisations began objecting to changing the NRTA and the Indian Act and, in the 1960s, several hunting rights cases, especially in the North, raised questions about the legality of proposed restrictions on Inuit and Indian hunting. There was increasing resentment and resistance at the local level, not only towards hunting restrictions as such, but also towards the methods of wildlife scientists. Both Inuit and Dene considered counting, tagging, and attaching radio or satellite collars, but most especially the matter of handling live animals, improper and disrespectful behaviour towards caribou. Caribou research methods soon came to be seen as the cause of caribou scarcity, rather than as an appropriate response to it.

In 1979–80 a third 'crisis' occurred, triggered by allegedly low herd counts and, again, sensational photos of large kills in Saskatchewan when wintering animals penetrated much farther south than usual. Again the same calls for immediate restrictions on aboriginal harvesting were heard, although in the outcome it transpired that the counts were incorrect and that there were in fact more caribou than the wildlife scientists had claimed. That episode, however, triggered the formation of the Beverly-Qamanirjuaq Caribou Management Board (and in a larger context, the wildlife chapters of the modern land claim agreements that provide for co-management and the priority of the aboriginal subsistence harvest). It also led to a recognition that the Inuit may have been right after all.

Acknowledgements

The work on which this paper is based was part of a project to compare caribou management systems in Alaska and Canada, funded by the United States Man and the Biosphere Program. I wish to acknowledge especially the support and ideas of J.K. Kruse, who led the project, and E.S. Burch, Jr., whose project in Alaska paralleled mine in Canada. I am grateful to F.J. Tough for undertaking archival research in support of this project, and to the Prince Albert Grand Council for permission to use material I obtained in the course of land claims research on behalf of the Saskatchewan Denesuline.

Notes

1   Csonka reports that the police almost never visited the section of tundra where Inuit and Dene met, and that there were only a few white trappers in the region (1999: 134).

2   The Northwest Game Act of 1917 was the applicable federal statute until 1949, when it was replaced by the NWT Game Ordinance. Both were administered by the Department of the Interior and its successors, and regulations were frequently amended on the advice of the Advisory Board on Wildlife Protection (ABWLP) and the Northwest Territories Council (the latter a mostly appointed body during the period).

3   These included a game laws paragraph that made Indians subject to provincial game laws 'provided, however, that the said Indians shall have the right ... of hunting, trapping and fishing game and fish for food at all seasons of the year on all unoccupied Crown lands'.

4   When the ABWLP was terminated in 1957, Barren Ground caribou management was coordinated by the federal-provincial administrative and technical committees recently established for that purpose.

5   NAC, RG22/7/33, RG22/249/40-6-6(2).

6   PC2567, 3 July 1947.

7   NAC, RG22/248/40-6-3(1), Gibson to Trocellier, 28 June 1950.

8   NAC, RG22/248/40-6-3(1), Trocellier to Young, 28 February 1951.

9   NAC, RG22/248/40-6-3(1), Sinclair to Commissioner, 5 December 1951.

10 NAC, RG22/248/40-6-3(2), Chief Alexis J.M. Beaulieu, St. Joseph's Hospital, Fort Resolution, to Young 15 November 1953; R.G. Robertson to Beaulieu, 30 November 1953.

11 NAC, RG22/96/32-2-5(2,3).

12 NWT Game Ordinance, 25.2.b.

13 NWT Game Ordinance, 26.2.a; NAC, RG22/248/40-6-3(2), Cunningham to Commissioner, RCMP, 8 June 1954.

14 NAC, RG22/248/40-6-3(2), Cunningham to Laviolette, secretary, Indian and Eskimo Welfare Commission, University of Ottawa, 9 June 1954.

15 NAC, RG22/248/40-6-3(2), André Renaud to Jean Lesage, 6 August 1954.

16 NAC, RG22/248/40-6-3(2), Cunningham to DM, 7 June 1955.

17 NAC, RG22/4/14, ABWLP Minutes, 7 February 1938.

18 NAC, RG22/1/14.

19 PC2567, 3 July 1947.

20 NAC, RG85/148/400-11-12(3).

21 Sometimes luridly described, e.g. 'There are times when aborigines take caribou at a disadvantage and are carried away by the lust of killing, so that they are quite beside themselves and slaughter far more caribou than they and their families and friends can possibly use.' (NAC, RG22/248/40-6-3(1), Gibson to Trocellier, 28 June 1950.)

22 NAC, RG22/96/32-2-5(3).

23 NAC, RG22/96/32-2-5(3).

24 NAC, RG22/16/69, Richards paper on wildlife resources in the NWT, to 14th Provincial-Dominion Wildlife Conference, 1950.

25 NAC, RG22/96/32-2-5(3), ABWLP Minutes, 17 March 1950.

26 NAC, RG85/360/3-1-6-7-1-A(4), Burton to Fraser, 1 November 1954.

27 NAC, RG22/248/40-6-3(2), Sivertz to Commissioner, 23 August. 1955.

28 NAC, RG22/248/40-6-3(2).

29 NAC, RG22/248/40-6-3(2), Churchman to Robertson, 29 August 1955.

30 NAC, RG85/1944/A401-22(1).

31 NAC, RG10/8933/140/20-16(2), McGilp memo, 20 May 1965.

32 NAC, RG85/1944/A401-22(1).

33 NAC, RG22/4/14.

34 NAC, RG22/96/32-2-5(3).

35 NAC, RG22/96/32-2-5(3), Meeting re caribou, 2 February 1950.

36 NAC, RG22/96/32-2-5(3), ABWLP Minutes, 17 March 1950. In an appendix to the minutes of this meeting, I. McT. Cowan recommended that limits be placed on Native killing wherever these could be enforced, noting, however, that unenforceable regulations bred harmful disrespect for authority.

37 NAC, RG22/96/32-2-5(3), ABWLP Minutes, 17 August 1950. Clancy (1983: 26–27) suggests that in practice, when Indian Affairs staff acted as JPs, they refused to convict. As well, the Act provided for exemptions to the close season when survival was at stake, which was, arguably, all of the time for Inuit and Dene on the eastern part of the range. This exemption gave considerable discretion to enforcement officers and at the same time made convictions difficult to secure.

38 NAC, RG85/636/420-2/101(1).

39 NAC, RG22/96/32-2-5(3), Meeting re caribou, 2 February 1950.

40 NAC, RG22/248/40-6-3(1), October 1950.

41 NAC, RG22/16/69, ABWLP, November 1950.

42 Cited in NAC, RG85/360/3-1-6-7-1-A(4), Cunningham to Fraser, 12 October 1954.

43 We found no record of charges laid, or convictions obtained, against Denesuline or Inuit on the eastern part of the caribou range, under either territorial or provincial wildlife ordinances, during the period under review. Weather station employees at Ennadai were found to have killed caribou in violation of the NWT Game Ordinance in 1963, but due to the statute of limitations, no charges were laid (RG85/1944/A400-1(1)).

44 NAC, RG85/360/3-1-6-7-1-A(4), Memo by Kelsall on legislation and enforcement, 1954.

45 NAC, RG85/360/3-1-6-7-1-A(4), Burton to Fraser, 1 November 1954.

46 NAC, RG85/360/3-1-6-7-1-A(4), Fraser to Director, 5 November 1954.

47 NAC, RG22/213/40-6-6.

48 See, for example, P.X. Mandeville Patrol Reports from Fort Smith, 1957, NAC, RG85/636/420-2/101(1), and 1959, NAC, RG10/8406/601/20-10(1), and 12 June 1958, memo re Barren Ground caribou situation in the NWT, NAC, RG85/1944/A401-22(1).

49 NAC, RG85/148/400-11-12(3), Provincial Dominion Conference 1949.

50 NAC, RG10/8399/501/20-9-2 (1), patrol report, 22 November 1951.

51 NAC, RG10/8399/501/20-9-2 (1).

52 NAC, RG22/248/40-6-3(2).

53 See, for example, Csonka (1995: 368, fn.362), and for a more general discussion of 'ilira' (an Inuktitut word for a particular type of fear of people and their power) and its effects on Inuit, Brody (1975).

54 NAC, RG85/148/400-11-12(1).

55 NAC, RG85/148/400-11-12(1).

56 Apparently first proposed in unattributed memo of 12 June 1958, and recommended in 1960 (NAC, RG85/1944/A401-22(1), Bolger to Brown, 23 August 1960).

57 PC1960-1256. These events are described in more detail by Clancy (1987) and Cranstonsmith (1995).

58 NAC, RG13/2723/19000-1.

59 The Inuit relocations in the Keewatin are described in detail in Tester and Kulchyski (1994) and Marcus (1995). It seems probable that at least some famine episodes were induced by sedentarisation and reduced mobility of the Inuit, rather than any actual disappearance or unavailability of caribou as such.

60 One thing that was not done, however, was to implement a programme of floor prices or subsidies for fur (Clancy 1983).

61 In a popular account of the history of caribou research, Urquhart observed: 'By today's standards none of the survey results would be considered even remotely accurate, much less meriting comparison with each other to establish a trend. But with aerial censusing being "high tech" for that era, the data were accepted with few reservations' (Urquhart 1989: 100).

12 Epilogue: Cultivating Arctic Landscapes

Mark Nuttall

The contributors to this volume have discussed the relationship of circumpolar peoples to the wild animals which support them, and the politics of regulation of Arctic landscapes, hunting and herding in a range of local cultural contexts. Although the ethnographic material ranges across a vast part of the northern reaches of the globe, the contributors point to the compelling similarities in the political and cultural settings of aboriginal peoples, together with their common experiences about how various capitalist and socialist states claimed control over their lands and animals. The majority of contributors describe how animals such as caribou and reindeer are the cultural, economic and often spiritual foundation for many northern peoples, including the Gwich'in of Alaska and northwestern Canada, the Saami of Fennoscandia and Russia's Kola Peninsula, and numerous peoples of Siberia and the Russian Far East, such as the Evenki and Chukchi. Inuit groups in Alaska, Canada and Greenland, while mainly hunters of marine mammals, often see caribou meat as an important subsistence food or an important source of income. Arctic peoples identify closely with the caribou they hunt and the reindeer they herd. They are dependent on them for much of their food, and they use their hides for clothing, for tents and other shelters. Yet these animals not only sustain indigenous peoples in an economic sense; they provide a fundamental basis for social identity, cultural survival and spiritual life, which is illustrated by rich mythologies, vivid oral histories, festivals and ceremonies.

While the contributors to this book focus mainly on the relationships between people and caribou and reindeer, and the institutional settings in which hunting and herding occur, the chapters as a whole deal with a range of critical issues of pressing contemporary concern throughout the circumpolar North. Their arguments and ethnographies have far-ranging implications and suggest some general conclusions. They illustrate the contested perspectives on how Arctic landscapes should be cultivated and managed, how people who depend on animals for survival work within (or are constrained by) new and evolving forms of management and governance, and how hunters and herders work with scientists and policy makers – often claiming the right to set policy themselves on the basis of their knowledge about animals and Arctic ecosystems. Reindeer herding and caribou hunting also throw into relief some generic global problems, provoking debate about who has the right to make decisions about nature and wildlife.

Some of the most pressing issues facing indigenous peoples are those concerning the impacts on reindeer and caribou, and local economies and cultures, of resource development, climate change, rights to manage and use herds, and questions of title to land. Environmental problems arising from industrial activity both outside and within the Arctic have serious consequences not only for traditional livelihoods, but also for human health and the health of animal populations. In addition, local economies are vulnerable to changes caused by the global processes affecting markets, technologies and public policies. All this poses challenges to hunters and herders – especially in the ways their relationships to the environment and animals are transformed.

Arctic hunters and herders have always lived with and adapted to shifts and changes in the size, distribution, range and availability of animal populations. They have dealt with flux and change by developing significant flexibility in resource procurement techniques and in social organisation. Yet the ecological and social relations between indigenous peoples and animal species are not just affected by climate-induced disruption, changing habitats and migration routes, or new technology. The livelihoods of the indigenous peoples of the Arctic are subject both to the ebb and flow of the market economy and to the implementation of government policy that either contributes to a redefinition of hunting, herding and fishing, or threatens to subvert subsistence lifestyles and indigenous ideologies of human–animal relationships. Hunters and herders are concerned and anxious about industrial development, resource exploitation and environmental degradation because of the close relationship between the cultural, economic, political and ecological situations of indigenous communities in the circumpolar North.

Throughout the circumpolar North, indigenous peoples live under greatly circumscribed conditions, the majority of them in permanent settlements with elaborate infrastructures, and their hunting, herding and fishing activities are determined to a large extent by resource management regimes and local and global markets. Various contributors to this book show how changes to settlement patterns and the ecological relations between humans and animals arise from government

attempts to introduce new economic activities or to sedentarise indigenous peoples. Gray's chapter describes how the Soviet authorities
'industrialised' reindeer herding as a way of facilitating the development of the Soviet North. The new settlements and industries in
Siberia came to depend on reindeer herders to supply them with meat.
However, with the collapse of the Soviet Union came the disappearance of what was, in effect, a largely artificial and subsidised market
for reindeer meat. The immediate post-Soviet period, in which a new
capitalist society began to emerge, was characterised by confusion in
deciding on ownership of animals and title to land and migration
routes. The increase in the sizes of herds, together with the encroach-
ment of industrial development, has pushed reindeer herders onto
smaller tracts of land and restricted migration routes for their herds,
with the result that some areas of pasture are overgrazed and degraded.
Today, privatisation and the transition to a market economy bring new
challenges to reindeer herding peoples in Siberia and the Russian Far
East, highlighting the dependence of Arctic reindeer systems on the
complex interlinkages between local, regional and global economies.
In a similar vein, Usher argues that caribou management on the Canadian Barren Grounds became an integral part of a broad programme of
social engineering – federal, provincial and territorial authorities
imposed management strategies based on their own (rather than Inuit
and Dene) ideas about conservation and hunting. There are similar stories from other parts of the circumpolar North – Dau (2000), for example, argues that the introduction of reindeer to the Seward Peninsula in
western Alaska during 1892–1902 was done to provide meat for Inupiat communities, yet was also intended as a way of transforming Inupiat from being subsistence marine mammal hunters to reindeer
herders and thus playing an active role in the wider cash economy of
the United States.

   Although governments have been keen to industrialise, commer-
cialise and control reindeer herding and caribou hunting, strict regulatory regimes and management practices imposed by states and
federal and provincial agencies actually act to hinder this process.
While aiming, in principle, to protect and conserve wildlife they also
restrict access to resources. In Alaska, for example, state and federal
policies make subsistence issues extremely complex to say the least –
with state and federal law defining subsistence as the customary and
traditional noncommercial use of wild resources, regulation limits the
prospects of finding markets for meat and fish. Earning money through
more commercial channels is not an option for Alaskan subsistence
hunters. In northern Fennoscandia, Saami ways of life have been tied

to reindeer herding, hunting and fishing, and, because Saami have never been organised at a social and political level into different tribal groupings, the northern parts of Norway, Sweden and Finland have not been divided into firm Saami territorial boundaries. Nomadic Saami reindeer herders have traditionally ranged far and wide, crossing national borders as they follow their reindeer herds between winter and summer pastures. In modern times, political developments have restricted migration routes over the last hundred years or so. Economic development in the nineteenth and twentieth centuries, such as mining, forestry, railways, roads, hydroelectric power and tourism have all had their impact on traditional Saami livelihoods.

In Greenland, threats to the cultural and economic viability of hunting livelihoods in small communities come from transformations in resource management regimes and Home Rule government regulations, which conflict with local customary practices and knowledge systems (Dahl 2000; Nuttall 2001). Caribou, whales, seals and fish, which have traditionally been subject to common-use rights vested in members of a local community, are becoming national and privately owned divisible commodities subject to rational management regimes defined by the state and the interest groups of hunters and fishers, rather than locally understood and worked-out rights, obligations and practices. As is still evident in some parts of Greenland today, it has traditionally been the case that no one owns animals – everyone has the right to hunt and fish as a member of a local community. A caribou, fish or sea mammal does not become a commodity until it has been caught and transformed into private property. Even then, complex local rules, beliefs and cultural practices counter the exclusive sense of individual ownership (Nuttall 2001: 67). However, trends in caribou hunting since the 1980s are illustrative of general wildlife management policies in Greenland (discussed at length in Sejersen's chapter), where membership of a territorial, or place-based community no longer gives hunters exclusive rights to harvest caribou. In west Greenland, caribou hunting was largely a family event until the 1970s. Kinship, locality and territory were the mechanisms for regulating harvesting activities. Today, hunting rights are vested in people as members of social and economic associations irrespective of a local focus. Discussing the situation in central west Greenland, Dahl (2000) shows how the traditional hunting territories of various communities are not the same as the administrative boundaries that surround villages, towns, districts and municipalities. The relevant territorial unit for hunting caribou (and other animals such as beluga and narwhal) is Greenland, rather than a place-based community. In this new administrative domain, the knowledge that hunters and fishers possess relating to animals and fish differs from the scientific models put forward by biologists as the basis for management. For hunters and fishers, marine mammals, fish and caribou are living beings – to catch them, one must be respectful, skilful and knowledgeable, and understand their precise movements as individuals in an environment that is shifting constantly. For biologists, animals and fish belong to the abstract totality of the stock and can be measured, counted and managed accordingly.

Reindeer herders, caribou hunters and marine mammal hunters are thus constrained by institutional frameworks and management structures and are commonly experiencing a transition from herding and hunting (from what we may call a 'way of life') to an occupation and industry. This is apparent in the chapters by Bjørklund, Beach and Wishart. The similarities with fisheries management in the circumpolar North are notable, especially the effects of the implementation of individual transferable quotas (ITQs). The ITQ system is a management response to overfishing and declining catches of major fish species, particularly demersal fish. Although designed to ensure the viability of fish stocks, sustainable catch levels, and economic efficiency, ITQ management results in the transformation of traditional common-use rights in fish stocks into privately owned, divisible commodities. As Helgasson and Palsson (1997) argue, ITQs represent the idea that both the human and natural worlds can be organised, controlled and managed in a rational way. Nature is not only 'presented as an inherently technical and logical domain, the project of the resource economist and manager is sometimes likened to that of the engineer or the technician' (ibid.: 452). Helgasson and Palsson describe the public discontent in Iceland with the commoditisation of fishing rights as a consequence of the ITQ system, a system which has resulted in fishing rights being concentrated in the hands of a few large operators – a discontent articulated in feudal metaphors such as 'tenancy' and 'lordships of the sea'. The ITQ system, although ostensibly seen by economists and resource managers as a way of achieving the sustainable use of fish stocks, has in reality a social impact in terms of changing power relations within local communities and regional fisheries, by contributing to the concentration of wealth in the hands of a few large fishing vessel owners. The ITQ system has effectively meant the enclosure of the commons and the privatisation of resources, which allows parallels to be drawn between fisheries and rural land-use debates throughout the Arctic.

One consequence of new forms of management is that hunters, herders and fishers, people who live from the land and the sea, run the risk of being vilified and labelled eco-criminals. In the 1980s and 1990s, animal-rights groups and environmentalist organisations mobilised international public opinion and spurred political action against the harvesting of marine mammals and fur-bearing animals by indigenous peoples in the Arctic. Many were not only opposed to hunting on moral grounds, but also expressed concern at what they considered to be the development of 'nontraditional' activities, such as markets for seal meat and whale meat (Wenzel 1991). In this volume, Usher discusses reports about Inuit and Dene engaging in 'wanton slaughter', 'excessive kills' and 'needless waste'. Today similar reports of overharvesting and indiscriminate killing of wildlife, as well as wasting meat, by indigenous peoples in parts of the Arctic echo the concerns of a previous generation of biologists, are catching the public eye and are attracting considerable media attention (see Hansen 2002), and hunting is increasingly seen as a morally suspect and even criminal activity which contributes to the subversion of existing traditional social orders. This perception of the hunter, herder or fisher as someone engaging in morally suspect practices, and even becoming progressively deskilled, does nothing to advance the claim put forward by indigenous peoples for governments, managers and policy makers to recognise the legitimacy of indigenous knowledge for informing new ways of thinking about appropriate forms of wildlife management.

Caribou are seen by environmentalists, wildlife managers and tourists (see Wishart's contribution) as symbols of wild nature under threat. Caribou are either overharvested (as in Greenland), or are threatened by big business and global change (oil development, pollution). Indigenous peoples are either the culprits, overharvesting wildlife, or the saviours of the environment and the people who really know how to manage and look after the herds. As Beach and Bjørklund mention in this volume, reindeer are not only major economic determinants to the livelihood of indigenous Saami pastoralists in Sweden and Norway, but are also powerful symbols, moving contemporary debates of resource conflict dramatically into larger discourses concerning the difficult relations between a native minority and the nation-state and between environmentalists and the Saami herders, who have increasingly been cast in terms of eco-criminals.

In Alaska and Canada, most caribou gather in large herds of tens of thousands to more than one hundred thousand animals on their calving grounds in the brief Arctic summer, and scatter widely in small groups for the rest of the year. In both indigenous accounts and the accounts of others (scientists, environmentalists) caribou and reindeer are inextricably linked to Arctic landscapes – landscapes which, for indigenous peoples, are part of an all-encompassing homeland, and for scientists and environmentalists are natural and wild. Popular images and representations of thousands of caribou ranging over the vast tundra landscapes of Canada and Alaska also inform popular perceptions and fuel the expectations of tourists. As Wishart shows, tourists driving along the highways on northern Canada expect to see an abundance of Arctic animals moving across untrammelled wilderness, and complain that it is not worthwhile travelling to the Arctic if they do not see any. Tourists need to be able to view wildlife in a situation that fulfils their expectations of pristine wilderness. In this sense, their expectations are that the Arctic is a wildlife park in which they can travel safely and gaze upon majestic herds of grazing animals.

Increasingly, reindeer herders and caribou hunters (along with other resource harvesters, such as Norwegian whalers and Greenlandic seal hunters) perceive themselves to be victims rather than beneficiaries of wildlife management and resource-use policy. Highly competing and contested interests over who has rights to manage and exploit resources are at stake throughout the circumpolar North, together with arguments that many animal populations are at unsustainable levels. Arctic hunting and herding societies not only face tremendous challenges to their ways of life, but also to reclaiming their rights. They respond to these challenges by questioning scientific expertise and official assumptions about what should count as valid knowledge for informing wildlife management. Hunters and herders accuse scientists and policy makers of acting with insufficient consideration about the workings of the ecosystem and are calling for a greater involvement in research and survey projects, as well as in the actual processes of negotiating and implementing management systems.

It becomes clear in all the chapters comprising the present volume that there is a fundamental clash between a dominant form of scientific knowledge and the situated knowledge of local people. As Kofinas et al. (2000) point out, although modern science has greatly expanded understanding of the relationships among Arctic species, there has been little appreciation of the value of local and traditional knowledge in understanding these systems. The contributors in this book underline this argument – Usher argues that officials in Canada had poor knowledge of caribou herds and their distribution and migration patterns, but ignored Inuit and Dene views, while Nagy shows how, on Banks Island, governments and wildlife managers chose to let the muskox population increase while the Inuvialuit wanted to keep it at a low level, fearing they would drive the caribou away (which, in the event, happened). Nagy argues how government policies aimed at the protection of one species – muskox – ignored the concerns of the Inuvialuit, who should have been allowed to participate in decisions concerning the control of the muskox population in order to keep a sustainable population of caribou for their subsistence.

Wildlife management in the circumpolar North has tended to focus on catch and harvest, with the ultimate goal of stock management. Increasingly such management is exemplified by regimes that invest property rights in a resource and are designed to ensure the economic efficiency of hunting and herding. Yet, one optimistic conclusion arising from the ethnographic material presented here is that indigenous peoples are reclaiming their rights to be involved in the design and implementation of wildlife management. In Greenland, Canada and Alaska, Inuit groups have over two decades of experience in developing and putting into practice their own environmental strategies and policies to safeguard the future of Inuit resource use. These strive to ensure a workable participatory approach between indigenous peoples, scientists and policy-makers to the sustainable management and development of resources.

Co-management and participatory approaches to development and environmental management reject simplistic models that make marked distinctions between human settlement and the natural environment and focus instead on how human knowledge of the environment is actually constructed and used as a foundation upon which decisions relating to the effective local management of natural resources are made. As is often stated in the literature on the subject, the argument put forward for the co-management of wildlife, for example, is that by integrating indigenous knowledge with scientific knowledge, the resulting resource management system is better informed and suited to the resource, the people who rely on it, and to the needs of scientists and conservationists.

Co-management and participatory approaches to resource use and environmental management are not always successful, however. Discussing the relationship between environmental problems and human rights, Beach (1997) and Collings (1997a) examine, respectively, how Saami reindeer herding as a livelihood is threatened by both state rationalisation and environmentalists in Sweden, and the political and cultural contexts of wildlife management in Canada. In the Saami case, reindeer herding is under pressure from legislation to make it conform to the Swedish state's view of what constitutes profitable business rather than understanding it from a Saami cultural and economic perspective; while environmentalists criticise Saami reindeer herders for abandoning what they see as a 'traditional' lifestyle. For northern

Canada, Collings examines the virtues and drawbacks of co-management of wildlife and, although the indigenous environmental knowledge of Inuit hunters is considered 'useful' by scientists, Collings argues that such knowledge is not always taken as valid in the same way as the scientific knowledge of biologists. Rather than knowledge being transferred and shared, it is more often controlled, leading to a situation of the passive involvement of local communities rather than active involvement. Thus full community participation is not necessarily achieved, calling into question the effectiveness of co-management as both process and policy.

The value of indigenous knowledge and the need to rethink wildlife management has gained greater international recognition throughout the Arctic in recent years. The work of indigenous peoples' organisations has been critically important in this regard (Cruikshank, this volume; Nuttall 2000). Since the 1980s, indigenous peoples' organisations have become increasingly important actors in Arctic environmental politics, giving a greater voice to indigenous peoples throughout the circumpolar North and arguing the case for the inclusion of indigenous knowledge in strategies for environmental management and sustainable development. Over the last decade in particular, these organisations have played a pivotal role in agenda setting and political debate with respect to the Arctic environment and resource development, and have gained international visibility and credibility through their participation in policy dialogue and decision-making processes at regional, national and international levels. The main organisations, namely the Inuit Circumpolar Conference (ICC), the Saami Council, and the Russian Association of Indigenous Peoples of the North (RAIPON), have set themselves in the vanguard of environmental protection. They are now major players on the stage of international diplomacy and policy making concerning the future of the Arctic, which is apparent in the increasingly influential way they contribute to discussions in the Arctic Council. The Inuit Circumpolar Conference, in particular, has been a driving force behind many recent initiatives in Arctic environmental protection and sustainable development.

As Thorpe describes for Nunavut, federal and territorial authorities are working with hunters and devising new ways of incorporating local knowledge into wildlife management and environmental governance. Indeed, Nunavut government legislation requires biologists and wildlife mangers to recognise the importance of indigenous knowledge about animals and the environment. Not all governments in the Arctic are as progressive. As Sejersen points out for Greenland, biologists there are not yet convinced that local knowledge has any useful part to play in managing animals. As the Nunavut case shows, evolving forms of environmental governance in the North American Arctic include broader social institutions, everyday local knowledge, rules and practices which provide the framework for decision making and cooperation, rather than the formal organisations which exist in order to address and deal with environmental issues. A governance approach to wildlife management rejects the current instrumental emphasis and recognises that while existing wildlife management policies may have their weaknesses, one reason for their failure also lies in the administrative and political institutional contexts within which those policies and schemes arise and are legitimated. A governance approach takes a much more encompassing approach to wildlife management by looking at the institutional contexts, including indigenous peoples at all levels, and also by considering the wider social, economic, political and market conditions.

What emerges from the papers collected here is that, by adopting a governance approach to conceptualise new ways of thinking about the sustainable uses of reindeer and caribou, the focus of the research and policy-making agenda is widened to include consideration of the complex interlinkages between people, animals and different levels of the ecosystem; the assessment of pollution and climate change trends; the production and supply of reindeer and caribou meat for human food; and the interrelationships between harvesting, processing, marketing and consuming reindeer and caribou products. The chapters overall argue that there is an urgent need to recognise that there are broader aspects to managing animals than the pragmatic, rational, scientific, economic and instrumental dimensions which usually preoccupy resource managers and policy makers. A greater understanding is necessary of the social and cultural dynamics of reindeer herding and caribou hunting societies, the histories of wildlife management in various parts of the Arctic, the knowledge people have about the movement and behaviour of animals, and the social fabric of local communities and the networks of association in which resource users move and operate, if constructive dialogue is to take place between hunters and herders, industry managers, policy developers, environmentalists and others. This collection of papers is offered as a vital contribution to this dialogue.

Notes on Contributors

David G. Anderson is Senior Lecturer in Social Anthropology at the University of Aberdeen. His work focuses on the identity, political ecology and social movements of rural and aboriginal peoples living in the circumpolar Arctic. His ongoing collaborative research is with Evenki, Dolgan and Sakha reindeer hunters and herders in eastern Siberia and with Gwich'in and Inuvialuit hunters in Canada's western Arctic. His theoretical interests extend to the sociology of post-socialist states, national identity and development economics. He is the author of *Identity and Ecology in Arctic Siberia* and co-editor of several volumes including *Ethnographies of Conservation: Environmentalism and the Distribution of Privilege* (2003).

Hugh Beach is Professor of Cultural Anthropology at Uppsala University, Sweden. His primary field of expertise concerns Saami reindeer pastoralism in Sweden and Saami issues in general, although his broad involvement in the rights and livelihoods of northern peoples has brought him to field research in Norway, Canada, Alaska, northern China and Russia. His research project work includes the effects of the Chernobyl nuclear disaster for the Saami, a survey of northern living conditions, post-Soviet socioeconomic transformations among northern Russian indigenous peoples, and the politics of wildlife management and the threat of "ecolonialism" in the North.

Ivar Bjørklund is an anthropologist based at Tromso Museum, Norway. His areas of research interest are ethnic identity and resource management in northern Norway. He has many years' experience of working on ecological knowledge, systems of tenure and the social organisation of reindeer herding in Saami coastal and inland rural areas. His current research includes the social and economic impact of changes to the regulation of access to marine resources and the transformation of reindeer herding.

Craig Campbell is a doctoral research student in the Department of Sociology at the University of Alberta. His research involves the use of still photography, video technology, and digital media to record perceptions of the land among Evenki reindeer pastoralists in Siberia. He is the principle curator of the Internet-based photo-gallery 'revealing pictures & reflexive frames'.

Julie Cruikshank is Professor of Anthropology at the University of British Columbia, Canada. Her research focuses on practical and theoretical developments in oral tradition studies; specifically, how competing forms of knowledge become enmeshed in struggles for legitimacy. Her ethnographic experience is rooted in the Yukon Territory, where she lived and worked for many years recording life stories with Athapaskan and Tlingit elders. She has also carried out comparative research in Alaska and Siberia. Her current work draws on theoretical trends linking the anthropology of memory with environmental anthropology. She is author of *Life Lived Like a Story* (1990), written in collaboration with three Yukon elders, Angela Sidney, Annie Ned and Kitty Smith, *Reading Voices* (1991) and *The Social Life of Stories* (1998).

Patty A. Gray received her PhD in Cultural Anthropology from the University of Wisconsin-Madison in 1998. She is a Research Fellow in the Siberian Project Group at the Max Planck Institute for Social Anthropology in Halle, Germany, and Assistant Professor of Anthropology at the University of Alaska Fairbanks. Her continuing research concerns social movements, transformation in rural communities, and regional political struggle in rural Russia and the Russian North. She is author of *Indigenous Activism in the Russian Far North: The Chukotka Case* (forthcoming).

Murielle Nagy has an MA in Archaeology from Simon Fraser University and a PhD in Anthropology from the University of Alberta. Since 1990 she has coordinated three major oral history projects for the Inuvialuit of the western Canadian Arctic. In 1999 she participated in the preparation of an oral history project for the Vuntut Gwich'in of Old Crow in the Yukon. She has been awarded various postdoctoral fellowships and research grants to work at GÉTIC of the Université Laval on the anthropological research of Oblate missionary Émile Petitot, who lived among the Dene and the Inuvialuit of northwest Canada from 1862 to 1881. She was the coordinator of IASSA (International Arctic Social Sciences Association) from 1998 to 2001, editorial assistant to the journal *Études/Inuit/Studies* from 1997 to 2002, and is now its editor.

Mark Nuttall is Professor of Anthropology at the University of Alberta, Canada. He specialises in the anthropology of rural and coastal communities in the Arctic and North Atlantic and has carried out research in Greenland, Alaska, Scotland and Canada. His work has a particular focus on local knowledge, marine mammal hunting and local fisheries, and the social impact of climate change. He is author of *Arctic Homeland: Kinship, Community and Development in Northwest Greenland* (1992), editor of *The Encyclopaedia of the Arctic* (Routledge, forthcoming) and co-editor of several works.

Frank Sejersen has an MA in Anthropology and a PhD from the Faculty of Humanities at the University of Copenhagen. He is Associate Professor at the Department of Eskimology in Copenhagen. Since the end of the 1980s he has been working on the human dimensions of resource management, indigenous peoples' politics and issues of self-determination, primarily in the Arctic. He has done fieldwork in Greenland in the town of Sisimiut, and is currently working on sustainability, local knowledge and landscape visions in Greenland. In 1998 he organised the Third International Congress of Arctic Social Sciences.

Natasha Thorpe has worked and travelled with Inuit throughout the Kitikmeot region of the Canadian Arctic for the last seven years. She holds a Masters degree in Resource Management from Simon Fraser University in Vancouver. Her most recent work was as principal researcher for the Tuktu and Nogak project (TNP), a community-driven effort to document and communicate Inuit knowledge of caribou and calving grounds for the Bathurst caribou herd. She is co-author (with Naikak Hakongak, Sandra Eyegetok and Kitikmeot Elders) of *Thunder on the Tundra: Inuit Qaujimajatuqangit of the Bathurst Caribou* (2003). She currently works with Golder Associates Ltd. in Victoria (Canada).

Peter J. Usher is an Ottawa-based consultant who specialises in resource management, impact assessment and aboriginal claims, and has written extensively on these topics. A graduate in geography from McGill University and the University of British Columbia, he has over forty years of research and practical experience in the Arctic and sub-Arctic regions of Canada.

Robert P. Wishart is the Northern Studies Centre post-doctoral fellow, University of Aberdeen. His current research is on the legacy of Scottish contacts with First Nations peoples in Canada and its imapct on how landscapes are perceived. He has also written on the ethnohistory and persistence of hunting practices among Ojibwe and Potawatomi in Southwestern Ontario, Canada.

Bibliography

Adams, Marie, Kathryn J. Frost, and Lois A. Harwood. 1993. 'Alaska and Inuvialuit Bel-
  uga Whale Committee (AIBWC) – an Initiative in "at home" management', *Arctic* 46(2):
  134–137.
Agrawal, A. 1995. 'Dismantling the Divide Between Indigenous and Scientific Knowl-
  edge'. *Development and Change* vol. 26: 413–39.
Akana, J. 1998. Elder in Umingmaktuuq, Nunavut. Interview by Eileen Kakolak, Doris
  Keyok, and Natasha Thorpe, 8 June. Tape recording. Tuktu and Nogak Project, Iqaluk-
  tuuttiaq.
Algona, B. 1999. Hunter in Qurluqtuq, Nunatvut. Interview by Sandra Eyegetok and
  Natasha Thorpe, 2 November. Tape recording. Tuktu and Nogak Project, Iqaluktuuttiaq.
Algona, M. 1999. Elder in Qurluqtuq, Nunatvut. Interview by Sandra Eyegetok and
  Natasha Thorpe, 1 November. Qurluqtuq, Nunavut. Tape recording. Tuktu and Nogak
  Project, Iqaluktuuttiaq.
Alonak, J. 1998. Elder at Huiqqittaaq River, Nunavut. Interview by Sandra Eyegetok and
  Natasha Thorpe, 8 August. Tape recording. Tuktu and Nogak Project, Iqaluktuuttiaq.
Alvaraz, R.D. and J.A. Diemer. 1998. 'The Democratization of Science: Toward a Produc-
  tive Synthesis of Common and Scientific Knowledge'. *Bridging Traditional Ecological
  Knowledge and Ecosystem Science: Conference Proceedings.* Compiled by Ronald L.
  Trosper. 13–15 August, 1998. Northern Arizona University. Flagstaff, Arizona: 24–35
  and at Povungnituk, Quebec. 2 vols. Inukjuak, Quebec: Avataq Cultural Institute, 1983.
Anderson, David G. 1991. 'Turning Hunters into Herders: A Critical Examination of Soviet
  Development Policy Among the Evenki of South-eastern Siberia', *Arctic*, 44(2): 12–22.
Anderson, David G. 2000a. *Identity and Ecology in Arctic Siberia: The Number One Rein-
  deer Brigade.* Oxford: Oxford University Press.
Anderson, David G. 2000b. 'Rangifer and Human Interests', *Rangifer* 20, (2–3): 153–74.
Anderson, David G. (ed.) 2001. *Narodnaia Meditsina.* Novosibirsk: Sibprint.
Anderson, David G., and Eeva Berglund (eds.) 2003. *Ethnographies of Conservation: Envi-
  ronmentalism and the Distribution of Privilege.* Oxford: Berghahn.
Anonymous. 1982. 'History of caribou research', *Caribou News* 2(4): 4–5.
Anonymous. 1998a. 'Fiskere og fangere skal tages med på råd', *Sermitsiaq*, 16. October, p.
  26.
Anonymous. 1998b. 'Hudbremsefluen dræber rensdyrene', *Sermitsiaq*, 14 August, p. 7.
Anonymous. 1998c. 'Tvivlsom teori om rensdyrdød', *Sermitsiaq*, 4 September, p. 11.
Anonymous. 1998d . Interview by Natasha Thorpe, 3 June. Qingauk, Nunavut. Tape
  recording. Tuktu and Nogak Project, Iqaluktuuttiaq.
Anonymous. 1999. 'Fangere og biologer i samarbejde om rensdyr', *Atuagagdliutut/Grøn-
  landsposten*, 22 July, p. 11.
Archival Material: File N92–091 from the NWT Archives, Yellowknife (NWT).
Atatahak, G. 1999. Researcher, Naonayaotit Traditional Knowledge Study: Qurluqtuq,
  Nunavut. Email to N. Thorpe.

Atatahak, G. 2000. Researcher, Naonayaotit Traditional Knowledge Study: Qurluqtuq, NU. Email to N. Thorpe.

Banfield, A.W.F. 1954. *Preliminary Investigation of the Barren-Ground Caribou.* Canadian Wildlife Service Wildlife Management Bulletin, Series 1, 10A and 10B.

Banfield, A.W.F. 1956. 'The Caribou Crisis', *The Beaver*, Outfit 286: 3–7.

Banfield, A.W.F. 1961a. *A Revision of the Reindeer and Caribou Genus Rangifer. Biological Series Bulletin*, 66. Ottawa: National Museums of Canada.

Banfield, A.W.F. 1961b. 'Migrating Caribou', *Natural History* 20: 56–64.

Barnaby, J. 1987. 'University Research and the Dené Nation'. Summary of a presentation to the ACUNS meetings in Yellowknife, in *Education, Research, Information Systems and the North*, ed. W.P. Adams, Ottawa: Association of Canadian Universities for Northern Studies.

Barr, William. 1991. *Back from the Brink: Road to Muskox Conservation in the Northwest Territories.* Komatik Series, Number 3. The Arctic Institute of North America, University of Calgary, Calgary.

Barth, Fredrik. 1995. 'Other Knowledge and Other Ways of Knowing'. *Journal of Anthropological Research*, 51: 65–68.

Basso, Keith. 1990. *Western Apache Language and Culture.* Tucson: University of Arizona Press.

BCMPC (Bathurst Caribou Management Planning Committee). 2000. Bathurst Caribou Management Meeting. Cambridge Bay, NU. Enekniget Katimayet Centre, 25–27 April.

Beach, Hugh. 1981. *Reindeer-Herd Management in Transition: The Case of Tuorpon Saameby in Northern Sweden*, Acta Univ. Ups., Uppsala Studies in Cultural Anthropology 3. Uppsala.

Beach, Hugh. 1993. 'Straining at Gnats and Swallowing Reindeer: the Politics of Ethnicity and Environmentalism in Northern Sweden', in *Green Arguments and Local Subsistence*, ed. G. Dahl. Stockholm Studies in Social Anthropology.

Beach, Hugh. 1994. 'Shots Heard Round the World', in *Beslutet om Småviltjakten – En studie i myndighetsutövning*, ed. Agneta Arnesson-Westerdahl. Utgiven av Sametinget.

Beach, Hugh. 1997 'Negotiating Nature in Swedish Lapland: Ecology and Economics of Saami Reindeer Management', in *Contested Arctic*, ed. E.A. Smith and J. McCarter. Seattle: University of Washington Press.

Beach, Hugh. 2000a. 'The Saami' in *Endangered Peoples of the Arctic: Struggles to Survive and Thrive*, ed. Milton Freeman. Endangered Peoples of the World Series. London: Greenwood Press.

Beach, Hugh. 2000b. 'Reindeer-Pastoralism Politics in Sweden: Protecting the Environment and Designing the Herder', in *Negotiating Nature*, ed. Alf Hornborg and Gísli Pálsson. Lund Studies in Human Ecology 2. Lund University Press.

Beach, Hugh. 2001. 'World Heritage and Indigenous Peoples – the Example of Laponia', in *Upholders of Culture Past and Present*, ed. Bo Sundin. Royal Swedish Academy of Engineering Sciences (Kungl. Ingenjörsvetenskapsakademien–IVA), Stockholm: Elanders Gotab.

Berger, Thomas R. 1977. *Northern Frontier, Northern Homeland: The Report of the Mackenzie Valley Pipeline Inquiry*, 2 vols. Ottawa: Minister of Supply and Services.

Berger, Thomas R. 1985. *Village Journey: The Report of the Alaska Native Review Commission.* New York: Hill & Wang.

Bergerud, A.T. 1988. 'Caribou, Wolves and Man', *Trends in Ecology and Evolution* 3: 68–72.

Berkes, F. 1999. *Sacred Ecology: Traditional Ecological Knowledge and Resource Management.* Philadelphia: Taylor and Francis.

Berkes, F., A. Hughes, P.J. George, R.J. Preston, B.D. Cummins, and J. Turner. 1995. 'The Persistence of Aboriginal Land Use: Fish and Wildlife Harvest Areas in the Hudson and James Bay Lowland, Ontario', *Arctic* 48(1): 81–93.

Berry, J.W. 1999. 'Aboriginal Cultural Identity', *The Canadian Journal of Native Studies* XIX(1): 1–36.

Bjerre, Michael, and Karl Erik Nielsen. 1998a. 'Han vil sende danskerne hjem', *Berlingske Tidende*, 8 June, p. 6.

Bjerre, Michael, and Karl Erik Nielsen. 1998b. 'Ungdomsoprøret i Nuuk'. *Berlingske Tidende*, 7 June, pp. 2–5.

Bjørklund, I. 1988. 'For mye rein i Finnmark?', *Finnmark Dagblad*. 28 December.

Bjørklund, I. and T. Brantenberg. 1981. *Samisk reindrift – norske inngrep*. Universitetforlaget.

Bjørklund, I. and H. Eidheim. 1999. 'Om reinmerker – kulturelle sammenhenger og norsk jus i Sapmi', in *Norsk ressursforvaltning og samiske rettighetsforhold*, ed. I.Bjørklund. Ad Notam Gyldendal

Bjørklund, I. 1990. 'Sami Reindeer Pastoralism as an Indigenous Resource Management System in Northern Norway – a Contribution to the Common Property Debate', *Development and Change*. SAGE, London, Newbury Park and New Delhi, 2: 75–86.

Blehr, O. 1964. 'Action Groups in a Society with Bilateral Kinship: A Case Study from the Faroe Islands', *Ethnology* 2: 269–75.

Bogoras, Waldemar. 1904–09. *The Chukchee*. New York: American Museum of Natural History Memoirs, 11 vols. Volume 7, Parts 1–3, of the Jesup North Pacific Expedition Publications.

Borgmann, Albert. 1984. *Technology and the Character of Contemporary Life*. Chicago: Chicago University Press.

Bourque, J. 1981. 'Wildlife Management in the Northwest Territories', in *Fish, Fur and Game for the Future: A Summary of Presentations and Discussions*. Conference proceedings from 23–25 February 1981. Yellowknife: Science Advisory Board of the Northwest Territories. Working Paper No. 4: 9.

Bradshaw, Corey J.A., and Daryll M. Hebert. 1996. 'Woodland Caribou Population Decline in Alberta: Fact or Fiction?' *Rangifer*, special issue 9(16): 223–33.

Brice-Bennett, Carol, Alan Cooke, and Nina Davis (eds.) 1977. *Our Footprints are Everywhere: Inuit Land Use and Occupancy in Labrador*. Nain: Labrador Inuit Association.

Briggs, J.L. 1970. *Never in Anger: Portrait of an Eskimo Family*. Cambridge, MA: Harvard University Press.

Brightman, Robert. 1993. *Grateful Prey: Rock Cree Human-Animal Relationships*. Berkeley: University of California Press.

Brody, Hugh. 1975. *The Peoples' Land*. Toronto: Penguin.

Brody, Hugh. 1981. *Maps and Dreams*. Vancouver: Douglas and McIntyre.

Bromley, B. 2001. Personal communication to N. Thorpe. Yellowknife, NWT.

Brøsted, Jens. 1986. 'Territorial Rights in Greenland: Some Preliminary Notes'. *Arctic Anthropology* 23(1–2): 325–38.

Brush, S.B. 1993. 'Indigenous Knowledge of Biological Resources and Intellectual Property Rights', *American Anthropologist* 95: 653–86.

Bruun, O., and A. Kalland. 1995. *Asian Perceptions of Nature: A Critical Approach*. Richmond, UK: Curzon Press.

Burch, E.S., Jr. 1994. 'Rationality and Resource Use Among Hunters', in *Circumpolar Religion and Ecology: An Anthropology of the North*. ed. T. Irimoto and T. Yamada. Tokyo: University of Tokyo Press. pp. 163–85.

Burch, E.S., Jr. 1995. 'Caribou Management in Northwestern Alaska: Cultural and Historical Perspectives'. Unpublished draft report submitted to the Man and the Biosphere High Latitudes Ecosystem Directorate. Camp Hill, Pennsylvania.

Calef, G. 1981. *Caribou and the Barren-Lands*. Toronto: Firefly Books.

Canadian Wildlife Service. No date [1951]. *A Question of Survival: The Barren Ground Caribou*. Ottawa: Minister of Public Works and Government Services.

Canadian Wildlife Service. 2000. Hinterland Who's Who – Caribou, Web page, [accessed September 2001]. URL: http://www.cws-scf.ec.gc.ca/hww-fap/caribou/caribou.html.

Catton, Theodore. 1997. *Inhabited Wilderness: Indians, Eskimos, and National Parks in Alaska*. Albuquerque: University of New Mexico Press.

Childers, Robert, and Mary Kancewick. nd. 'The Gwich'in (Kutchin): Conservation and Cultural Protection in the Arctic Borderlands'. Unpublished manuscript prepared for The Gwich'in Steering Committee.

Cizek, P. 1990. *The Beverly-Kaminuriak Caribou Management Board: A Case Study of Aboriginal Participation in Resource Management*. Ottawa: Canadian Arctic Resources Committee Background Papers 1.

Clancy, P. 1983. 'Game Policy in the Northwest Territories: The Shaping of Economic Position'. Paper presented at the Annual Meeting of the Canadian Political Science Association, Vancouver B.C.

Clancy, P. 1987. 'Native Hunters and the State: The "Caribou Crisis" in the Northwest Territories', *Studies in National and International Development* Occasional Paper no. 87–101. Kingston: Queen's University.

Clarke, Simon. 1992. 'The Quagmire of Privatization', *New Left Review* 196: 3–28.

Clausen, B. 1981. 'Hunting and Country Food Programs in Greenland', in *Fish, Fur and Game for the Future*. Conference proceedings from 23–25 February 1981. Yellowknife: Science Advisory Board of the Northwest Territories. Working Paper No. 4: 9.

Cohen, D.W. 1989. 'The Undefining of Oral Tradition', *Ethnohistory* 36(1): 9–18.

Collings, Peter. 1997a. 'The Cultural Context of Wildlife Management in the Canadian North', in Contested Arctic: *Indigenous Peoples, Industrial States and the Circumpolar Environment*, ed. Eric Alden Smith and Joan McCarter. Seattle: University of Washington Press, pp. 13–40.

Collings, Peter. 1997b. 'Subsistence Hunting and Wildlife Management in the Central Canadian Arctic', *Arctic Anthropology* 34(1): 41–56.

Condon, R. 1995. 'The Rise of the Leisure class: Adolescence and Recreational Acculturation in the Canadian Arctic', *Ethos* 23(1): 47–68.

Condon, R. 1996. *The Northern Copper Inuit: A History*. Toronto: University of Toronto Press.

Condon, R., P. Collings, and G. Wenzel. 1995. 'The Best Part of Life: Subsistence Hunting, Ethnicity, and Economic Adaptation Among Young Adult Inuit Males'. *Arctic* 48(1): 31–46.

Connerton, Paul. 1989. *How Societies Remember*. Cambridge: Cambridge University Press.

Council for Yukon Indians. 1991. *Voices of the Talking Circle: Yukon Aboriginal Languages Conference 1991*. Ottawa: Department of Secretary of State.

Couturier, S., Rehaume Courtois, Helene Crepeau, Louis-Paul Rivest, and Stuart Luttich. 1996. 'Calving Photocensus of the Riviere George Caribou Herd and Comparison with an Independent Census', *Rangifer*, special issue 9(16): 283–96.

Cramér, T. 1968–2002. *Samernas Vita Bok*. Stockholm.

Cranstonsmith, V. 1995. 'Chipewyan Hunting, Scientific Research and State Conservation of the Barren-Ground Caribou, 1940–1970'. Unpublished MA thesis in Native Studies. Saskatoon: University of Saskatchewan.

Cruikshank, Julie, in collaboration with A. Sidney, K. Smith and A. Ned. 1990. *Life Lived Like a Story: Life Stories of Three Yukon Elders.* Lincoln: University of Nebraska Press.

Cruikshank, Julie. 1990. 'Getting the Words Right: Perspectives on Naming and Places in Athapaskan Oral History', *Arctic Anthropology*, 27(1): 52–65.

Cruikshank, Julie. 1998. *The Social Life of Stories: Narrative and Knowledge in the Yukon Territory.* Lincoln: University of Nebraska Press.

Csonka, Y. 1995. *Les Ahiarmiut, a l'écart des Inuit Caribous.* Neuchâtel: Éditions Victor Attinger.

Csonka, Y. 1999. 'A Stereotype Further Dispelled: Inuit-Dené relations West of Hudson Bay, 1920–1956', *Inuit Studies* 23(1–2): 117–44.

Cunnison, Ian. 1951. History of the Luapula: An Essay on the Historical Notions of a Central African Tribe. London: Oxford University Press. Quoted in 'The Undefining of Oral Tradition', E.W. Cohen. *Ethnohistory* 36(1), 1989.

Dahl, Jens. 1986a. 'Greenland: Political Structure of Self-government', *Arctic Anthropology* 23(1–2): 315–24.

Dahl, Jens. 1986b. *Arktisk Selvstyre.* Copenhagen: Akademisk forlag.

Dahl, Jens. 1989. 'The Integrative and Cultural Role of Hunting and Subsistence in Greenland', *Études/Inuit/Studies* 13(1): 23–42.

Dahl, Jens. 1993. 'Indigenous Peoples of the Arctic'. Paper presented at the Nordic Council's Arctic Conference, Reykjavík 16–17 August.

Dahl, Jens. 1998. 'Resource Appropriation, Territories and Social Control', in *Aboriginal Environmental Knowledge in the North*, ed. Louis-Jacques Dorais, Murielle Nagy and Ludger Müller-Wille. Québec: GÉTIC, University of Laval, pp. 61–80.

Dahl, Jens. 2000. *Saqqaq. An Inuit Hunting Community in the Modern World.* Toronto: University of Toronto Press.

Dahlström, Esa N. 2003. 'Negotiating Wilderness in a Cultural Landscape. Predators and Saami Reindeer Herding in the Laponian World Heritage Area', ACTA Universitatis Upsaliensis. Uppsala Studies in Cultural Anthropology No.32 Uppsala: Upsaliensis Acadamiae.

Dau, J. 2000. 'Managing Reindeer and Wildlife on Alaska's Seward Peninsula', *Polar Research* 19(1): 57–62.

Davis, J. 1981. 'Caribou Management Programs in Alaska', in *Fish, Fur and Game for the Future.* Conference Proceedings from 23–25 February 1981. Yellowknife: Science Advisory Board of the Northwest Territories. Working Paper No. 4: 5.

Davis, S. and K. Ebbe. (eds.) 1995. *Traditional Knowledge and Sustainable Development: Proceedings of a Conference held at the World Bank, Washington, September 27–28.* Washington: Environmentally Sustainable Development Proceedings, Series No. 4.

De Coccola, Raymond and Paul King. 1986. 'The incredible Eskimo', in *Life among the Barren Land Eskimo*, Surrey: Hancock House.

DeFrancis, John. 1989. *Visible Speech: The Diverse Oneness of Writing Systems.* Honolulu: University of Hawaii Press.

Descola, P. and G. Palsson. 1996. *Nature and Society: Anthropological Perspectives.* New York: Routledge.

DIAND (Department of Indian Affairs and Northern Development). 1984. *Western Arctic Land Claim. Inuvialuit Final Claim Settlement.* DIAND, Ottawa.

Dick, Lyle. 2001. *Muskox Land. Ellesmere Island in the Age of Contact.* Calgary: University of Calgary Press.

Dikov, N.N. 1989. *Istoriia chukotki s drevneishikh vremen do nashikh dnei.* Moscow: Mysl'.

Diubaldo, R. 1985. *The Government of Canada and the Inuit: 1900–1967*. Ottawa: Research Branch, Corporate Policy, Indian and Northern Affairs Canada. QS-3413-000-EE-A1.

Dizard, Jan E. 1994. *Going Wild: Hunting, Animal Rights and the Contested Meaning of Nature*. Amherst: University of Massachusetts Press.

Dmytryshyn, Basil, E.A.P. Crownhart-Vaughan, and Thomas Vaughan (eds.) 1985. *Russia's Conquest of Siberia, 1558–1700: A Documentary Record*. Volume 1 of *To Siberia and Russian America: Three Centuries of Russian Eastward Expansion, 1558–1867*. Portland, Oregon: Oregon Historical Society.

Dorais, Louis-Jacques. 1996. 'Inuugatta inuulerpugut: Kalaallit and Canadian Inuit Identitites', *Cultural and Social Research in Greenland 95/96*, pp. 28–33.

Douglas, William O. 1964. 'Banks Island: Eskimo Life on the Polar Sea', *National Geographic* 125: 705–35.

Druri, Ivan Vasil'evich. 1989. 'Kak byl sozdan pervyi olenesovkhoz na Chukotke', *Kraevedcheskie Zapiski (Magadan)* XVI: 3–14.

Dybbroe, Susanne. 1989. 'Danske horisonter – og grønlandske: Advokaterne, eksperterne og den "indfødte" befolkning efter hjemmestyret', in *Dansk mental geografi. Danskernes syn på verden – og på sig selv*. ed. Ole Høiris. Aarhus: Aarhus Universitetsforlag, pp. 149–61.

Dybbroe, Susanne. 1996. 'Questions of Identity and Issues of Self-determination', *Études/Inuit/Studies* 20(2): 39–53.

Dybbroe, Susanne. 1999. 'Researching Knowledge: The Terms and Scope of a Current Debate', in *Changes in the Circumpolar North: Culture, Ethics and Self-determination*, ed. Frank Sejersen. Copenhagen: International Arctic Social Sciences Association, pp. 13–26.

Ellen, R. 1996. 'Introduction', in *Redefining Nature: Ecology, Culture and Domestication*. ed. R. Ellen and K. Fukio. Oxford: Berg, pp. 1–36.

Eyegetok, S. 1999. Senior Researcher, Tuktu and Nogak Project. Personal communication to N. Thorpe, Iqaluktuuttiaq.

Eyegetok, S. 2000. Senior Researcher, Tuktu and Nogak Project. Personal communication to N. Thorpe, Iqaluktuuttiaq.

Fehr, Alan and William Hurst. (eds.) 1997. *A Seminar on Two Different Ways of Knowing: Indigenous and Scientific Knowledge*. Inuvik: Aurora Research Institute.

Feit, Harvey. 1988. 'Self Management and State Management: Forms of Knowing and Managing Northern Wildlife', in *Traditional Knowledge and Renewable Resource Management in Northern Regions*, ed. M.M.R. Freeman and L.N. Carbyn. Edmonton, Alberta: Boreal Institute for Northern Studies, pp. 72–91.

Feit, Harvey. 1991. 'Gifts of the Land: Hunting Territories, Guaranteed Incomes, and the Construction of Social Relations in James Bay Cree Society', in *Cash, Commoditisation and Changing Foragers*. ed. N. Pearson and T. Matsuyama. Senri Ethnological Studies 30. Osaka: National Museum of Ethnology.

Feit, Harvey A. 1994. 'The Enduring Pursuit: Land, Time and Social Relationships in Anthropological Models of Hunter-Gatherers and in Subarctic Hunters' Images'. in *Key Issues in Hunter-Gatherer Research*, ed. Ernest S. Jr. Burch and Linda Ellanna. Oxford: Berg, pp. 421–40.

Feit, Harvey. 1998. 'Reflections on Local Knowledge and Wildlife Resource Management: Differences, Dominance and Decentralization', *Aboriginal Environmental Knowledge in the North*, ed. Louis-Jaques Dorais, Murielle Nagy and Ludger Muller-Wille. Quebec: Getic, pp. 123–48.

Ferguson, M. 2000. Regional Wildlife Biologist, Department of Sustainable Development, Government of Nunavut: Pond Inlet, Nunavut. Personal communication to N. Thorpe.

Ferguson, M., R.G. Williamson, and F. Messier. 1998. 'Inuit Knowledge of Long Term Changes in a Population of Arctic Tundra Caribou', *Arctic* 51(3): 201–19.

Fienup-Riordan, Ann. 1990. *Eskimo Essays: Yup'ik Lives and How We See Them*. New Brunswick and London: Rutgers University Press.

Fienup-Riordan, Ann. 1992. 'One Mind, Many Paths: Yup'ik Eskimo Efforts to Control Their Futures', *Études/Inuit/Studies* 16(1–2): 75–83.

Fienup-Riordan, Ann. 1994. *Boundaries and Passages: Rule and Ritual In Yup'ik Eskimo Oral Tradition*. Norman: University of Oklahoma Press.

Fienup-Riordan, Ann. 1997. 'Metaphors of Conversion, Metaphors of Change', *Arctic Anthropology* 34(1): 102–16.

Fienup-Riordan, Ann. 1999. 'Yaqulget qaillun pilartat (what the birds do): Yup'ik Eskimo Understanding of Geese and Those Who Study Them', *Arctic* 52(1): 1–22.

Flaherty, Martha. 1995. 'Freedom of Expression or Freedom of Exploitation', The *Northern Review* 14(summer): 178–85.

Fleischer, Jørgen. 1998. 'Politisk og administrativ udvikling i Grønland', in *Retsforhold ogsamfund i Grønland*, ed. Hanne Petersen and Jakob Janussen. Nuuk: Ilisimatusarfik, pp. 17–24.

Foigel, Isi. 1980. 'Home Rule in Greenland', Meddelelser om Grønland, (Man and Society), no. 1.

Fondahl, Gail. 1998. *Gaining Ground? Evenkis, Land, and Reform in Southeastern Siberia*. Boston: Allyn and Bacon.

Forchhammer, M.C. 1995. 'Sex, Age, and Season Variation in the Foraging Dynamics of Muskoxen, *Ovibus moschatus*, in Greenland', *Canadian Journal of Zoology* 73: 1344–61.

Forchhammer, Søren. 1992. 'Whaling in Upernavik Municipality'. Paper presented at the 8th Inuit Studies Conference, Laval University, Québec.

Freeman, M.M.R. 1985. 'Appeal to Tradition: Different Perspectives on Arctic Wildlife Management', in *Native Power: The Quest for Autonomy and Nationhood of Indigenous Peoples*, ed. J. Brosted, J. Dahl, A. Gray, H.C. Gullov, G. Henriksen, J.B. Jorgensen and I. Kleivan. Bergen: Universitetsforlaget AS, pp. 264–81.

Freeman, M.M.R. 1989. 'Graphs and Gaffs: a Cautionary Tale in the Common Property Resource Debate', *Common Property Resources: Ecology and Community-based Sustainable Development*, ed. E. Berkes. London: Belhaven Press, pp. 92–109.

Freeman, M.M.R. 1992a. 'Ethnoscience, Prevailing Science, and Arctic Co-operation', in *Arctic Alternatives: Civility or Militarism in the Circumpolar North*, ed. F. Griffiths. Toronto: Samuel Stevens, pp. 79–99.

Freeman, M.M.R. 1992b. 'The Nature and Utility of Traditional Ecological Knowledge', *Northern Perspectives* 20(1): 9–12.

Freeman, M.M.R., and Ludwig N. Carbyn, (eds.) 1988. *Traditional Knowledge and Renewable Resource Management in Northern Regions*. Edmonton: Boreal Institute for Northern Studies and IUCN Commission on Ecology.

Freeman, M.M.R. (ed.) 1976. *Inuit Land Use and Occupancy Project*. Ottawa: Indian and Northern Affairs.

Fuller, W.A., and B.A. Hubert. 1981. 'Fish, Fur and Game in the Northwest Territories: Some Problems of and Prospects for Increased Harvest', *International Symposium on Renewable Resources and the Economy of the North*, ed. M.M.R. Freeman. Ottawa: Association of Canadian Universities for Northern Studies, pp. 12–29.

GeoNorth Northern Environmental Consultants. 2002. *Final Report on the Aquatic Effects of the Ferry Landings at Tsiigehtchic and Fort McPherson*. Yellowknife: Government of the Northwest Territories, Department of Transportation.

George, Jane. 1996. 'Who Should Talk about Shamanism?' *Nunatsiaq News*, 23 August, pp. 11, 15.

Gessain, Robert. 1981. *Ovibos. La grande aventure des boeufs musqués et des hommes.* Paris: Éditions Robert Laffont.

Government of Norway. 1985. *Forskning, veiledning og utdanning i reindriften. Tilrådning om langtidsplan for reindriftens fagtjeneste 1985–1995.* ('Grue-utvalget'), pp. IV–21–31. Landbruksdepartementet, Oslo.

Gracheva, Galina N. 1983. *Traditsionnoe mirovozzrenie okhotnikov Taimyra.* Leningrad: Nauka.

Grønhaug, R. 1976. *Chayanovs regel. Lov og struktur.* University of Bergen.

Grønlands Landsråds Forhandlinger. 1961. Beretninger vedrørende Grønland, no. 2.

Gulløv, Hans Christian. 1979. 'Home Rule in Greenland', *Etudes/Inuit/Studies* 3(1): 131–42.

Gunn, A., G. Arlooktoo and D. Kaomayo. 1988. 'The Contribution of the Ecological Knowledge of Inuit to Wildlife Management in the Northwest Territories', in *Traditional Knowledge and Renewable Resource Management*, ed. Milton M.R. Freeman and Ludwig N. Carbyn. Edmonton: The Canadian Circumpolar Institute, IUCN Commission on Ecology, pp. 22–30.

Gunn, Anne, Chris Shank, and Bruce McLean. 1991. 'The History, Status and Management of Muskoxen on Banks Island', *Arctic* 44(3): 188–95.

Guthrie, R. Dale. 2001. 'Origin and Causes of the Mammoth Steppe: a Story of Cloud Cover, Woolly Mammal Tooth Pits, Buckles, and Inside-out Beringia', *Quaternary Science Reviews* 20: 549–74.

Gwich'in Tribal Council, and Indian and Northern Affairs Canada. 1992. *Gwich'in Comprehensive Land Claim Agreement.* Ottawa: Indian and Northern Affairs Canada.

Hakongak, N. 1998. Interview by Natasha Thorpe and Meyok Omilgoetok, 11 May. Tape recording. Tuktu and Nogak Project, Iqaluktuuttiaq.

Hakongak, N. 2000. Wildlife Officer, Government of Nunavut: Vancouver, BC. Personal communication to N. Thorpe.

Hakongak, N. 2001. Wildlife Officer, Government of Nunavut: Vancouver, BC. Personal communication to N. Thorpe.

Haller, Albert A. 1986. *The Spatial Organization of the Marine Hunting Culture in the Upernavik District, Greenland.* Bamberg: University of Bamberg.

Hansen, B.V. (ed.) 1994. *Arctic Environment: Report on the Seminar on Integration of Indigenous Peoples' Knowledge.* Copenhagen: Ministry for the Environment (Iceland), Ministry for the Environment (Denmark), Home Rule of Greenland (Denmark Office).

Hansen, K. 2002. *A Farewell to Greenland's Nature.* Copenhagen: Gads Forlag.

Hardin, Garrett. 1968. 'The Tragedy of the Commons', *Science*, new series, 162(3859). (13 December 1968): 1243–48.

Harper, F. 1955. *The Barren-ground Caribou of Keewatin.* University of Kansas Museum of Natural History, miscellaneous publication 6.

Harper, K. 2000. 'Inuit Writing Systems in Nunavut', in *Nunavut: Inuit Regain Control of Their Lands and Their Lives*, ed. J. Dahl, J. Hicks and P. Jull. Copenhagen: International Work Group for International Affairs. Document 102, pp. 154–69.

Hearne, Samuel. 1968 [1795]. *Journey from Fort Prince Wales, in Hudson's Bay, to the Northern Ocean.* Philadelphia: Joseph & James Crukshank.

Heide-Jørgensen, Mads Peter. 1998. 'Et bæredygtigt Grønland – fornuftig forvaltning set gennem biologens briller', in *Seminar om de levende ressourcer, Katuaq 9–11 oktober 1998*, ed. Kirsten Rydahl and Ivalo Egede. Nuuk: Grønlands Naturinstitut, pp. 133–37.

Helgasson, A. and G. Palsson. 1997. 'Contested Commodities: the Moral Landscape of Modernist Regimes', *Journal of the Royal Anthropological Institute* 3: 451–71.

Hicks, J., and G. White. 2000. 'Nunavut: Inuit Self-determination Through a Land Claims and Public Government', in *Nunavut: Inuit Regain Control of Their Lands and Their Lives*, ed. J. Dahl, J. Hicks and P. Jull. Copenhagen: International Work Group for International Affairs. Document 102, pp. 30–117.

Hobson, G. 1992. 'Traditional Knowledge is Science', *Northern Perspectives* 20(1): 2.

Howard, A. and F. Widdowson. 1996. 'Traditional Knowledge Threatens Environmental Assessment', *Policy Options* vol. 17, November: 34–36.

Howard, L., R. Goodwin, and L. Howard (eds.) 1994. 'Indigenous Knowledge in Northern Canada: Annotated Bibliography'. Draft Prepared for the Canadian Polar Commission by Arctic Science Technology Information Systems, Arctic Institute of North America, University of Calgary.

Huber, T. and P. Pederson. 1997. 'Meteorological Knowledge and Environmental Ideas in Traditional and Modern Societies: The Case of Tibet', *Journal of the Royal Anthropological Institute (N.S.)* vol. 3: 577–98.

Humphrey, Caroline. 1995. 'Introduction' (to the special issue titled Surviving the Transition: Development Concerns in the Post-Socialist World), *Cambridge Anthropology* 18(2): 1–12.

Humphrey, Caroline. 1998. *Marx Went Away – But Karl Stayed Behind*. Ann Arbor: University of Michigan Press.

Hunn, E. 1999. 'The Value of Subsistence for the Future of the World', in *Ethnoecology: Situated knowledge/Located Lives*, ed. V.D. Nazarea. Arizona: University of Arizona Press, pp. 23–36.

Hunt, C. 1976. 'The Development and Decline of Northern Conservation Reserves', *Contact* 8(4): 30–75.

Huntington, Henry. 1989. 'The Alaska Eskimo Whaling Commission: Effective Local Management of a Subsistence Resource'. Thesis, Scott Polar Research Institute.

Huntington, Henry. 1992. *Wildlife Management and Subsistence Hunting in Alaska*. London: Belhaven Press in association with Scott Polar Research Institute.

Huntington, Henry. 1996. 'Recommendations on the Integration of Two Ways of Knowing: Traditional Indigenous Knowledge and Scientific Knowledge', in *Proceedings of the Seminar on the Documentation and Application of Indigenous Knowledge*. Inuvik, NWT, 15–17 November. (http://www.grida.no/caff/inuvTEK.htm)

Ihse, M. 1995. 'Renskötseln hot mot fjällnaturen: samerna kan inte fortsätta skylla allt på turisterna', in *Samefolket*, nr. 8: 14–15.

Inglis, J.T., (ed.) 1993. *Traditional Ecological Knowledge: Concepts and Cases*. Ottawa: Canadian Museum of Nature.

Ingold, Tim. 1980. *Hunters, Pastoralists and Ranchers: Reindeer Economies and Their Transformations*. Cambridge: Cambridge University Press.

Ingold, Tim. 1992. 'Culture and the Perception of the Environment', in *Bush Base, Forest Farm: Culture, Environment and Development*, ed. D. Parkin and E. Croll. London: Routledge, pp. 39–56.

Ingold, Tim. 1993. 'Globes and Spheres: The Topology of Environmentalism', in *Environmentalism: The View from Anthropology*, ed. Kay Milton. London: Routledge.

Ingold, Tim. 1996. 'Hunting and Gathering as Ways of Perceiving the Environment', in *Redefining Nature: Ecology, Culture and Domestication*, ed. R. Ellen and K. Fukio. Oxford: Berg, pp. 117–55.

Ingold, Tim. 2000. *The Perception of the Environment: Essays in Livelihood, Dwelling and Skill*. Routledge, London.

Ingold, Tim. 2002. Epilogue, in *People and the Land: Pathways to Reform in Post-Soviet Siberia*, ed. Erich Kasten. Berlin: Dietrich Reimer Verlag, pp. 245–55.

Inuit Circumpolar Conference. 1992. *Principles and Elements for a Comprehensive Arctic Policy*. Montreal: Centre for Northern Studies and Research.

Ipellie, A. 1997. 'Thirsty for Life: A Nomad Learns to Write and Draw', in *Echoing Silence: Essays on Arctic Narrative*, ed. J. Moss. Ottawa: University of Ottawa Press, pp. 93–101.

Irlbacher, S. 1997. 'The Use of Traditional Knowledge in Public Government Programs and Services in the Northwest Territories', Thesis (MA), University of Alberta.

Isham, James. 1949. *James Isham's Observations on Hudson's Bay, 1743 and Notes and Observations on a Book Entitled, 'A Voyage to Hudsons Bay in the Dobbs Galley, 1749'*, ed. E.E. Rich and A.M. Johnson. Toronto: Publ. Champlain Soc., Hudson's Bay Company, ser. 12.

Jaktlag SFS 1987:259. Utfärdad 1987-05-14, Stockholm.

Jaktförordning SFS 1987:905. Utfärdad 1987-09-24, Stockholm.

Jedrej, Charles and Mark Nuttall. 1996. *White Settlers. The Impact of Rural Repopulation in Scotland*. London: Gordon and Breach Science Publishers.

Jenness, Diamond. 1959. *People of the Twilight*. Chicago: University of Chicago Press.

Jenness, Diamond. 1967. *Eskimo Administration: IV. Greenland*. Montreal: Arctic Institute of North America.

Johansson, S. and N-G. Lundgren. 1998. 'Vad kostar en ren? En ekonomisk och politisk analys.' Rapport till expertgruppen för studier ioffentlig ekonomi. Ds 1998:8 Regeringskansliet, Stockholm.

Johnson, M. (ed.) 1992. *Lore: Capturing Traditional Ecological Knowledge*. Ottawa: Dené Cultural Centre and International Development Research Centre.

Jørgensen, Carl. 1964. 'Den administrative udvikling i teoretisk belysning', in *Grønland i udvikling*, ed. Guldborg Chemnitz and Verner Goldschmidt. København: Danmarks Radios Grundbøger, Fremad, pp. 65–124.

Kadlun-Jones, M. 2000. Language Consultant. Kadlun Services: Edmonton, AB. Personal communication to N. Thorpe.

Kadlun-Jones, M. 2001. Language Consultant. Kadlun Services: Edmonton, AB. Personal communication, email to N. Thorpe.

Kamoayok, L., 1998. Elder. Interview by Natasha Thorpe and Eileen Kakolak, 9 August, Hiukkittaak. Tape recording. Tuktu and Nogak Project, Iqaluktuuttiaq.

Kaniak, M. 1998. Elder. Interview by Natasha Thorpe and Eileen Kakolak, 9 August, Hiukkittaak. Tape recording. Tuktu and Nogak Project, Iqaluktuuttiaq.

Kay, C.E. 1994. 'Aboriginal Overkill: The Role of Native Americans in Structuring Western Ecosystems', *Human Nature* 5: 359–98.

Kelsall, J.P. 1968. *The Migratory Barren-ground Caribou of Canada*. Canadian Wildlife Service Monograph 3. Ottawa: Department of Indian Affairs and Northern Development.

Kendrick, A. 2000. 'Community Perceptions of the Beverly-Qamanirjuaq Caribou Management Board', *The Canadian Journal of Native Studies* XX, 1: 1–33.

Kevan, P.G. 1974. 'Peary Caribou and Muskoxen on Banks Island', *Arctic* 24: 256–64.

Kofinas, Gary P. 1998. 'The Costs of Power Sharing: Community Involvement in Canadian Porcupine Caribou Co-management'. Thesis (PhD), University of British Columbia.

Kofinas, Gary P., G. Osherenko, D. Klein and B. Forbes. 2000. 'Research Planning in the Face of Change: the Human Role in Reindeer/Caribou Systems', *Polar Research* 19(1): 3–21.

Krech, Shepard III. 1999. *The Ecological Indian: Myth and History*. New York: W.W. Norton and Company.

Kritsch, Ingrid. 1996. 'The Delta Science Camp', in *The Wild Times*, Fall issue, pp. 2–4. GNWT Department of Resources, Wildlife and Economic Development, NT.

Krupnik, Igor. 1993. *Arctic Adaptations: Native Whalers and Reindeer Herders of Northern Eurasia.* Hanover, NH: University Press of New England.

Krupnik, Igor. 2000a. 'Reindeer Pastoralism in Modern Siberia: Research and Survival in the Time of Crash', *Polar Research* 19(1): 49–56.

Krupnik, Igor. 2000b. *Pust' govoriat nashi stariki. Rasskazy aziatskikh eskimosov-iupik. Zapisi 1977–1987 gg.* Tula: Obshchestvo Yupik.

Krupnik, Igor, and H. Vakhtin. 1997. 'Indigenous Knowledge in Modern Culture: Siberian Yupik Ecological Legacy in Transition', *Arctic Anthropology* 34(1): 236–52.

Krupnik, Igor, and L. Bogoslovskaya. 1999. 'Old Records, New Stories: Ecosystem Variablity and Subsistence Hunting in the Bering Strait Area', *Arctic Research of the United States* 13: 15–24 .

Kuhn, R.G. and F. Duerden. 1996. 'A Review of Traditional Environmental Knowledge: An Interdisciplinary Canadian Perspective', *Culture* 16(1): 71–84.

Kulturrådet. 1990. *Kulturrådets betænkning 1990.* Nuuk: Grønlands Hjemmestyre.

Kuptana, G., Elder. 1998. Interview by Eileen Kakolak, Karen Kamoayok and Natasha Thorpe. 7 June, Umingmaktuuq. Tape recording. Tuktu and Nogak Project, Iqaluktuuttiaq.

Kuzyk, G.W., D.E. Russell, R.S. Farnell, R.M. Gotthardt, P.G. Hare and E. Blake. 1999. 'In Pursuit of Prehistoric Caribou on Thandlat, Southern Yukon', *Arctic* 52(2): 214–19.

Labrador Inuit Association. 1977. *Our Footprints are Everywhere: Inuit Land Use and Occupancy in Labrador.* Nain: Labrador Inuit Association.

Lakoff, G., and M. Johnson. 1980. *Metaphors We Live By.* Chicago: University of Chicago Press.

Langgaard, P. 1986. 'Modernisation and Traditional Interpersonal Relations in a Small Greenlandic Community: A Case study from Southern Greenland', *Arctic Anthropology* 23(1–2): 299–314.

Lantis, Margaret. 1972. 'Factionalism and Leadership: A Case Study of Nunivak Island', *Arctic Anthropology* 9(1): 43–65.

Larsson, Kjell. 2001. 'Rovdjuren som en del i vår biologiska mångfald – på gott och ont'. Presentation at the Annual Wolf Symposium in Vålådalen, Sweden, March 19–21, 2001.

Larter, Nicholas C. and John A. Nagy. 1997. 'Peary Caribou, Muskoxen and Banks Island Forage: Assessing Seasonal Diet Similarities'. Proceedings of the Second International Arctic Ungulate Conference, Fairbanks, Alaska, 13–17 August, 1995. *Rangifer* 17(1): 9–16.

Larter, Nicholas C. and John A. Nagy. 1999. 'Sex and Age Classification Surveys of Muskoxen on Banks Island, 1985–1990: A Review'. Manuscript report no. 113. Department of Resources, Wildlife and Economic Development, Government of the Northwest Territories, Inuvik.

Larter, Nicholas C. and John A. Nagy. 2000. 'Calf Production and OverWinter Survival Estimates for Peary Caribou, Rangifer Tarandus Pearyi, on Banks Island, Northwest Territories', *The Canadian Field-Naturalist* 114: 661–70.

Larter, Nicholas C. and John A. Nagy. 2001a. 'Variation Between Snow Conditions at Peary Caribou and Muskox Feeding Site and Elsewhere in Foraging Habitats on Banks Island in the Canadian High Arctic', *Arctic, Antarctic, and Alpine Research* 33(2): 123–30.

Larter, Nicholas C. and John A. Nagy. 2001b. 'Calf Production, Calf Survival, and Recruitment of Muskoxen on Banks Island During a Period of Changing Population Density from 1986–99', *Arctic* 54(4): 394–406.

Latour, Bruno. 1993. *We Have Never Been Modern.* Cambridge, MA: Harvard University Press.

Lawrie, A.H. 1948. *Barren Ground Caribou Survey, 1948.* Canadian Wildlife Service Report. C.873. Ms.

Légaré, A. 1998. 'An Assessment of Recent Political Development in Nunavut: The Challenges and Dilemmas of Inuit Self-government', *Canadian Journal of Native Studies* 18 (2): 271–99.

Lent, Peter C. 1999. *Muskoxen and Their Hunters: A History.* Animal Natural History Series, volume 5. Norman: University of Oklahoma Press.

Leont'ev, V.V. 1977. 'The Indigenous Peoples of Chukchi National Okrug: Population and Settlement', *Polar Geography* 1: 9–12.

Leont'ev, V. V., no date. *Osobennosti kul'turno-khoziaistvennogo razvitiia narodnostei Chukotki na sovremennom etape (1958–1967 gg.):* Shifr–Okonchatel'nyi otchet. Sibirskoe otdelenie AN SSSR severo-vostochnyi kompleksnyi nauchno-issledovatel'skii institut, laboratoriia arkheologii, istorii i etnografii.

Lewis, Henry T. 1982. *A Time for Burning: Traditional Indian Uses of Fire in the Western Canadian Boreal Forest.* Edmonton, Alberta: Boreal Institute for Northern Studies, University of Alberta.

Li, Tania M. 1996. 'Images of Community: Discourse and Strategy in Property Relations', *Development and Change* 27: 501–27.

Lidegaard, Mads. 1969. 'Grønlænderne og den moderne udvikling', in *Grønland i focus,* ed. Jan Hjarnø. København: Nationalmuseet, pp. 119–24.

Lindberg, C. 2001. 'Troféjakt på utrotningshotad art', in *Djurens Rätt!* no. 2, 2001.

Löfgren, N. 1999. 'Samer – kultur eller industri?' in Barometern-OT (Oskarshamnstidningen). *Insändare på insändarsidan* 99–06–25.

Lopez, Barry. 1986. *Arctic Dreams. Imagination and Desire in a Northern Lndscape.* Toronto: Bantam Books.

Lynge, F. 1992. *Arctic Wars, Animal Rights, Endangered People.* Hanover and London: Dartmouth.

Lynge, Finn. 1995. 'Indigenous Peoples Between Human Rights and Environmental Protection', *Nordic Journal of International Law* 64: 489–94.

Lynge, Ulla. 1991. 'Hvad med bygderne – blev planerne til virkelighed?' Thesis submitted at Ilisimatusarfik, Greenland.

Lyver, P., H. Moller and C. Thompson. 1999. 'Change in Sooty Shearwater *Puffinus Griseus* Chick Production and Harvest Precede ENSO Events', *Inter-Research,* Marine Ecology Progress Series. 188: 237–48.

Maghagak, A. 1999. Executive Secretary, Nunavut Tunngavik Incorporated, Iqaluktuuttiaq. Personal communication.

Maltin, E. 2000. 'Kikkaktok Science Camp, Walker Bay, Nunavut', *Above and Beyond* 12(3): 25.

Maniyogina, J. 1998. Interview by Sandra Eyegetok and Eva Komak. 10 July, Iqaluktuuttiaq. Tape recording. Tuktu and Nogak Project, Iqaluktuuttiaq.

Manning, T.H. and A.H. Macpherson. 1958. *The Mammals of Banks Island.* Arctic Institute of North America, Montreal.

Marcus, A. 1995. *Relocating Eden: The Image and Politics of Inuit Exile in the Canadian Arctic.* Hanover: University Press of New England.

Martin, Paul S. 1973. 'The Discovery of Amercia', *Science* 197: 969–74.

McDonald, M., L. Arragutainaq and Z. Novalinga. 1997. *Voices from the Bay: Traditional Ecological Knowledge of Inuit and Cree in the Hudson Bay Bioregion.* Ottawa: Canadian Arctic Resources Committee.

MacFarlane, R. 1905. *Notes on Mammals Collected and Observed in the Northern Mackenzie River District, Northwest Territories of Canada with Remarks on Explorers and Explorations of the Far North*. Proceedings (United States National Museum); vol. 28, no. 1405. Washington: US Government Printing Office.

McNabb, Steven. 1991. 'Elders, Iñupiat Ilitqusiat, and Culture Goals in Northwest Alaska', *Arctic Anthropology* 28(2): 63–76.

Meredith, Thomas C. 1983. 'Institutional Arrangements for the Management of the George River Caribou Herd: Remote or Local Control?' *Etudes/Inuit/Studies*, 7(2): 99–112.

Miller, Frank L. 1982. 'Caribou'. *Wild Mammals of North America: Biology, Management, and Economics*, ed. J. A. Chapman and G. A. Feldhamer. Baltimore: Johns Hopkins University Press, pp. 923–29.

Miller, Frank L. 1983. 'Restricted Caribou Harvest or Welfare – Northern Native's Dilemma', *Acta Zoologica Fennica* 175: 171–75.

Morehouse, Thomas. 1988. 'The Alaska Native Claims Settlement Act, 1991, and Tribal Government'. Institute of Social and Economic Research Occasional Papers, no. 19.

Morrow, P. and C. Hensel. 1992. 'Hidden Dimension: Minority-Majority Relationships and the Uses of Contested Terminology', *Arctic Anthropology* 29(1): 38–53.

Morrow, Phyllis. 1990. 'Symbolic Actions, Indirect Expressions: Limits to the Interpretation of Yupik Society', *Etudes/Inuit/Studies* 14(1–2): 141–58.

Muckenheim, Stephanie. 1998. *Importance of Fishing and Fish Harvesting to Yukon First Nations People: a Summary*. Whitehorse: Yukon Fish and Wildlife Management Board.

Muehlebach, Andrea. 2001. 'Making Place at the United Nations: Indigenous Cultural Politics at the UN Working Group of Indigenous Populations', *Cultural Anthropology* 16(3): 415–48.

Müller-Beck, Hansjürgen (ed.) 1977. 'Excavations at Umingmak on Banks Island, N.W.T., 1970 and 1973 Preliminary Report', in *Urgeschichtliche Materialhefte 1*, Institut für Urgeschichte der Universität Tübingen, Tübingen.

Nabokov, Peter. 2002. *Forest of Time: American Indian Ways of History*. Cambridge: Cambridge University Press.

Nadasdy, Paul. 1999. 'The Politics of TEK: Power and the "Integration" of Knowledge', *Arctic Anthropology* 36(1–2): 1–18.

Nagy, John A., Nicholas C. Larter and V.P. Fraser. 1996. 'Population Demography of Peary Caribou and Muskox on Banks Island, N.W.T., 1982–1992'. Proceedings of the Sixth North American Caribou Workshop, Prince George, British Columbia, Canada, 1–4 March 1994. *Rangifer*, special issue 9: 213–22.

Nagy, Murielle. 1999. 'Aulavik Oral History Project on Banks Island, NWT: Final Report'. Unpublished report prepared for the Inuvialuit Social Development Program, Inuvik.

Nagy, Murielle (ed.) 1999a. 'Aulavik Oral History Project: English Translations and Transcriptions of Interviews 3 to 30'. Unpublished manuscript prepared for the Inuvialuit Social Development Program, Inuvik.

Nagy, Murielle (ed.) 1999b. 'Aulavik Oral History Project: English Translations and Transcriptions of Interviews 31 to 72'. Unpublished manuscript prepared for the Inuvialuit Social Development Program, Inuvik.

Nagy, Murielle (ed.) 1999c. 'Aulavik Oral History Project: English Translations of Archival Tapes'. Unpublished manuscript prepared for the Inuvialuit Social Development Program, Inuvik.

Nakashima, D.J. 1986. 'Inuit Knowledge of the Ecology of the Common Eider in Northern Quebec', in *Eider Ducks in Canada*, ed. A. Reed. Ottawa: Canadian Wildlife Service, pp.102–13

Nakashima, D.J. 1990. *Application of Native Knowledge in EIA: Inuit, Eiders, and Hudson Bay Oil*. Canadian Environmental Assessment Research Council. Ottawa: Minister of Supply and Services.

Nakashima, D.J. 1993. 'Astute Observers on the Sea Ice Edge: Inuit Knowledge as a Basis for Arctic Co-management', in *Traditional Ecological Knowledge: Concepts and Cases*, ed. J.T. Inglis. Ottawa: Canadian Museum of Nature, pp. 99–110.

Neis, B., L.F. Felt, R.L. Haedrich and D.C. Schneider. 1999. 'An Interdisciplinary Method for Collecting and Integrating Fishers' Ecological Knowledge into Resource Management', in *Fishing Places, Fishing People*, ed. D. Newell and R.E. Ommer. Toronto: University of Toronto Press, pp. 217–38.

Nilsson, Eja. 1984. *Menneskenes land – 14 grønlændere fortæller*. København: Københavns Bogforlag.

Nooter, Gert. 1976. *Leadership and Headship. Changing Authority Patterns in an East Greenland Hunting Community*, trans. Seeger. Leiden: E.J. Brill.

Nordgrønlands Landsraadsforhandlinger. 1937. *Beretninger og Kundgørelser Vedrørende Styrelsen af Grønland*, 8: 1131–72.

Norrbottens Kuriren 1994 (newspaper of the northern Swedish province of Norrbotten), 9 December.

NSDC (Nunavut Social Development Council). 1999. Report on the Nunavut Traditional Knowledge Conference, Igloolik, 20–24 March 1988.

Nudds, Thomas D. 1988. 'Effects of Technology and Economics on the Foraging Behaviour of Modern Hunter-gatherer Societies', in *Knowing the North: Reflections on Tradition, Technology and Science*, ed. William C. Wonders. Edmonton, Alberta: Boreal Institute for Northern Studies, pp. 23–36.

Nuttall, Mark. 1992. *Arctic Homeland: Kinship, Community and Development in Northwest Greenland*. Toronto: University of Toronto Press.

Nuttall, Mark. 1998. *Protecting the Arctic: Indigenous Peoples and Cultural Survival*. Amsterdam: Harwood Academic Publishers.

Nuttall, M. 2000. 'Indigenous Peoples' Organisations and Arctic Environmental Co-operation', in *The Arctic: Environment, People, Policy*, eds. M. Nuttall and T.V. Callaghan. Amsterdam: Harwood Academic Publishers, pp. 621–37.

Nuttall, M. 2001. 'Locality, Identity and Memory in South Greenland', *Etudes/Inuit/Studies* 25(1–2): 53–72.

Oakes, Jill and Rick Riewe. 1996. 'Communicating Inuit Perspectives on Research', in *Issues in the North*, eds. Jill Oakes and Rick Riewe. Edmonton: Canadian Circumpolar Institute, pp. 71–80.

Okrainetz, G. 1992. 'Towards a Sustainable Future in Hudson Bay', *Northern Perspectives* 20(2): 12–16.

Oldendow, Knud. 1936. *Grønlands egne samfundsorganer*. København: Gads Forlag.

Olsen, Moses. 1969. Grønland – Danmark. 'Ensartedhed eller lighed?' in *Grønland i fokus*, ed. Jan Hjarnø. København: Nationalmuseet, pp. 81–92.

Orwell, George. 1945. *Animal Farm: A Fairy Story* . London: Secker & Warburg.

Osgood, Cornelius. 1970 [1936]. *Contributions to the Ethnography of the Kutchin*. New Haven: Human Relations Area Files Press.

Osherenko, Gail. 1988. *Sharing Power with Native Users: Co-Management Regimes for Arctic Wildlife*. Ottawa: Canadian Arctic Resources Committee.

Osherenko, Gail. 1988. 'Wildlife Management in the North American Arctic: The Case for Co-management,' in *Traditional Knowledge and Renewable Resource Management*, ed. M.M.R. Freeman and L.N. Carbyn. Edmonton: The Canadian Circumpolar Institute, IUCN Commission on Ecology, pp. 92–104.

Osherenko, Gail, and O.R. Young. 1989. *The Age of the Arctic: Hot Conflicts and Cold Realities*. Cambridge: Cambridge University Press.

Paine, Robert. 1964. 'Herding and Husbandry. Two Basic Concepts in the Analysis of Reindeer Management', *Folk* 6(1): 83–88.

Paine, Robert. 1970. 'Lappish Decisions, Partnerships, Information Management and Sanctions – a Nomadic Pastoral Adaption', *Ethnology* 9(1): 52–67.

Paine, Robert. 1994. *Herds of the Tundra: A Portrait of Saami Pastoralism.* Washington, DC: Smithsonian Institution Press.

Paine, R. 1982. *Dam a River, Damn a People?* Copenhagen: IWGIA.

Paine, Robert (ed.) 1977. *The White Arctic: Anthropological Essays on Tutelage and Ethnicity.* St. John's: Institute of Social and Economic Research.

Pálsson, Gísli. 1994. 'Enskillment at sea', *Man* 29(4): 901–27.

Parker, G.R. 1972. 'Biology of the Kaminuriak Population of Barren-Ground Caribou, part I: Total Numbers, Mortality, Recruitment, and Seasonal Distribution'. Canadian Wildlife Service Report Series, no. 20.

Pedersen, Sverre, Terry Haynes and Robert Wolfe. 1991. 'Historic and Current Use of Muskox by North Slope Residents, with Specific Reference to Kaktovik, Alaska'. Alaska Department of Fish and Game Division of Subsistence Technical Paper Series no. 206.

Pederson, Poul. 1995. 'Nature, Religion and Cultural Identity: The Religious Environmentalist Paradigm', in *Asian Perceptions of Nature: A Critical Approach*, ed. Ole Bruun and Arne Kalland. Richmond, UK: Curzon Press, pp. 258–76.

Pehrsson, R. 1957. *The Bilateral Network of Social Relations in Konkoma Lapp District.* Indiana University.

Petch, V. 1994. 'The Relocation of the Sayisi-Dene of Tadoule Lake'. Report prepared for the Royal Commission on Aboriginal Peoples.

Petersen, Didrik. 1994. 'Rejs hjem, hvis du ikke vil behandle os anstændigt', *Sermitsiak*, 26 August, p. 19.

Petersen, Robert. 1963. 'Family Ownership and Right of Disposition in Sukkertoppen District, West Greenland', *Folk*, 5: 269–81.

Petersen, Robert. 1965. 'Some Regulating Factors in the Hunting Life of Greenlanders', *Folk*, 7: 107–24.

Petersen, Robert. 1991. 'The Role of Research in the Construction of Greenlandic Identity', *North Atlantic Studies* 1(2): 17–22.

Petersen, Robert. 1993. 'Samfund uden overhovede – og dem med. Hvordan det traditionelle grønlandske samfund fungerede og hvordan det bl.a. påvirker fremtiden', *Grønlandsk Kultur- og Samfundsforskning* 93, 121–38.

Petersen, Robert. 1995. 'Colonialism as Seen From a Former Colonized Area', *Arctic Anthropology* 32(2): 118–26.

Petersen, Robert. 1998. 'Om ledelsesformer før og nu', in *Retsforhold og samfund i Grønland*, ed. Hanne Petersen and Jakob Janussen. Nuuk: Ilisimatusarfik, pp. 25–35.

Pika, Aleksandr. 1999. *Neotraditionalism in the Russian North: Indigenous Peoples and the Legacy of Perestroika.* Seattle: University of Washington Press.

Pitu. 1999. 'Lokal viden', *Pitu*, 1(1): 6.

Posey, Darrell. 1990. 'Intellectual Property Rights and Just Compensation for Indigenous Knowledge', *Anthropology Today* 6: 13–16.

Posey, D.A. and W.L. Overall (eds.) 1990. 'Ethnobiology: Implications and Applications'. *Proceedings of the First International Congress of Ethnobiology. Brazil.* 2 vols. Belem, Brazil: Museu Paraense Emilio Goeldi.

Povinelli, E. 1993. *Labor's Lot: The Power, History and Culture of Aboriginal Action.* Chicago: University of Chicago Press.

Povinelli, E. 1995. 'Do Rocks Listen? The Cultural Politics of Apprehending Australian Aboriginal Law', *American Anthropologist* 97(3): 505–18.

Preece, Rod. 1999. *Animals and Nature: Cultural Myths, Cultural Realities.* Vancouver: University of British Columbia Press

Quassa, Joan. 2001. Co-ordinator, Inuit Qaujimajatuqangit Working Group, Government of Nunavut: Iqaluit. Personal communication to N. Thorpe.

Raine, Peter A. 1998. 'Sharing Ecological Wisdom Through Dialogues Across Worldview Boundaries', in *Bridging Traditional Ecological Knowledge and Ecosystem Science: Conference Proceedings*. R.L. Trosper, compiler. 13–15 August, 1998. Northern Arizona University. Flagstaff, Arizona: 24–35.

Raven, Gary. 1996. 'Role of the Elders: Yesterday and Today', in *Issues in the North*, ed. Jill Oakes and Rick Riewe. Alberta: Canadian Circumpolar Institute, pp. 52–54.

Regeringens proposition 2000/01:57 'Sammanhållen rovdjurspolitik', *Överlämnad av regeringen till riksdagen den 21 dec, 2000*. Stockholm.

Reynolds, Patricia. 1989. 'Status of a Transplanted Muskox Population in Northeastern Alaska', *Canadian Journal of Zoology*. 67(5): A26–A30.

Richard, Pierre R. and D.G. Pike. 1993. 'Small Whale Co-management in the Eastern Canadian Arctic: A Case History and Analysis', *Arctic* 46(2): 138–43.

Richards, P. 1993. 'Cultivation: Knowledge or Performance?' in *An Anthropological Critique of Development: the Growth of Ignorance*, ed. M. Hobart. London and New York: Routledge, pp. 61–78.

Richardson, B. 1993a. *People of Terra Nullius: Betrayal and Rebirth in Aboriginal Canada*. Vancouver: Douglas & McIntyre.

Richardson. B. 1993b. 'Harvesting Traditional Knowledge', *Nature Canada* 22(4): 30–37.

Ridington, Robin. 1990. *Little Bit Know Something: Stories in a Language of Anthropology*. Toronto: Douglas and McIntyre.

Riedlinger, Dyanna and Fikret Berkes. 2001. 'Contributions of Traditional Knowledge to Understanding Climate Change in the Canadian Arctic', *Polar Record* 37(203): 315–28.

Riewe, R. and L. Gamble. 1988. 'The Inuit and Wildlife Management Today', in *Traditional Knowledge and Renewable Resource Management*, ed. M.M.R. Freeman and L.N. Carbyn. Edmonton: The Canadian Circumpolar Institute, IUCN Commission on Ecology, pp. 31–37.

Rival, L (ed.) 1998. *The Social Life of Trees: Anthropological Perspectives on Tree Symbolism*. Oxford and New York: Berg.

Robertson, R.G. 1961. 'The Future of the North', *North* 8(2): 1–13.

Roepstorff, Andreas. 2000. 'The Double Interface of Environmental Knowledge', in *Finding Our Sea Legs – Linking Fishery People and Their Knowledge with Science and Management*, ed. Barbara Neis and Lawrence Felt. St. John's: Institute of Social and Economic Research at Memorial University, pp. 163–88.

Rowell, J. 1989. 'Survey of Reproductive Tracts from Female Muskoxen Harvested on Banks Island, NWT', in P.F. Flood (ed.), *Proceedings of the Second International Muskox Symposium*, Saskatoon, Saskatchewan, Canada, 1–4 October, Ottawa.

Rushforth, Scott. 1992. 'The Legitimation of Beliefs in a Hunter-Gatherer Society – Bearlake Athapaskan Knowledge and Authority', *American Ethnologist* 19(3): 483–500.

Rushforth, Scott. 1994. 'Political Resistance in a Contemporary Hunter-Gatherer Society: More about Bearlake Athapaskan Knowledge and Authority', *American Ethnologist* 21(2): 335–52.

Russell, F. 1898. *Explorations in the Far North: Being the Report of an Expedition Under the Auspices of the University of Iowa During the Years 1892, '93, and '94*. Iowa: State University of Iowa.

Sandlos, John. 2001 'From the Outside Looking in: Aesthetics, Politics, and Wildlife Conservation in the Canadian North', *Environmental History* 6(1): 6–31.

Sara, M.N. 1990. 'Baddjeealuhuslahka ja boazudoallupolitihka', *Diedut* no.2, Sami Instituhtta, Guovdageaidnu.

Scott, Colin. 1979. 'Modes of Production and Guaranteed Annual Income in James Bay Cree Society'. Unpublished PhD. thesis, McGill University.

Scott, Colin. 1988. 'Property, Practice and Aboriginal Rights Among Quebec Cree Hunters', in *Hunters and Gatherers 2: Property, Power and Ideology*, ed. Tim Ingold. Oxford: Berg.

Scott, Colin. 1996. 'Science for the West, Myth for the Rest? The Case of James Bay Cree Knowledge Construction', in *Naked Science: Anthropological Inquiry into Boundaries, Power and Knowledge*, ed. L. Nader. London: Routledge, pp. 69–86.

Scott, Colin and Harvey Feit. 1992. *Income Security for Cree Hunters : Ecological, Social and Economic Effects*. Montreal: Conseil Québécois de la Recherche Sociale.

Sejersen, Frank. 1998. 'Strategies for Sustainability and Management of People: An Analysis of Hunting and Environmental Perceptions in Greenland with a Special Focus on Sisimiut'. Dissertation submitted to the Faculty of Humanities in partial fulfilment of the requirements for the PhD. degree at the Department of Eskimology, University of Copenhagen.

SEPA, Swedish Environmental Protection Agency. 2002, beslut 2002-03-11. Ansökan om tillstånd att Avliva eller flytta en varg. (Decision concerning the application for permission to kill or move a wolf).

Shanin, Teodor. 1987. 'Short Historical Outline of Peasant Studies', in *Peasants and Peasant Societies*, 2nd edn, ed. Teodor Shanin. Oxford: Basil Blackwell, pp. 467–75.

Shapcott, Catherine. 1989. 'Environmental Impact Assessment and Resource Management, a Haida Case Study: Implications for Native people of the North', *The Canadian Journal of Native Studies* IX (1): 55–83.

Sillitoe, Paul. 1998. 'The Development of Indigenous Knowledge: A New Applied Anthropology', *Current Anthropology* 39(2): 223–52.

Sissons, J. 1993. 'The Systemization of Tradition: Maori Culture as a Strategic Resource', *Oceania* 64(2): 97–116.

Sivdleq. 1999. 'Issue Containing Articles on the Old People's Conference in Sisimiut', *Sivdleq*, 10 June.

Smith, Timothy. 1989b. 'The Role of Bulls in Pioneering New Habitats in an Expanding Muskox Population on the Seward Peninsula, Alaska', *Canadian Journal of Zoology* 67(5): 1096–1101.

Smith, Timothy. 1989a. 'The Status of Muskoxen in Alaska', *Canadian Journal of Zoology* 67(5): A23–A25.

Smith, Timothy. 1994. 'Muskox'. Alaska Department of Fish and Game Wildlife Notebook Series, 1994.

Sonne, Birgitte. 1982. 'The Professional Ecstatic in his Social and Ritual Position', *Scripta Instituti Donneriani Aboensis* XI: 128–50.

Sørensen, Axel Kjær. 1983. *Danmark-Grønland i det 20. århundrede – en historisk oversigt*. København: Nyt Nordisk Forlag Arnold Busck.

Sørensen, Bo Wagner. 1994. 'Jagten på den indre grønlænder', *Kvinder, Køn og Forskning* 2: 53–68.

Spence, M. 1999. *Dispossessing the Wilderness: Indian Removal and the Making of National Parks*. Oxford: Oxford University Press.

Spink, J. 1969. 'Historic Eskimo Awareness of Past Changes in Sea Level', *Musk-Ox* 5: 37–40.

St. meld. nr 28. 1991–1992. *En baerekraftig reindrift*. Landbruksdepartementet, Oslo.

Stefansson, Vilhjalmur. 1921. *The Friendly Arctic*. New York: McMillan Co.

Stefansson, Vilhjalmur. 1924. *The Northward Course of the Empire*. New York: Harcourt.

Stern, D. 2000. Hunter: Iqaluktuuttiaq. Personal communication to N. Thorpe.

Stevenson, M.G. 1996. 'Indigenous Knowledge in Environmental Assessment', *Arctic* 49(3): 278–91.

Stevenson, M.G. 1997. 'Ignorance and Prejudice Threaten Environmental Assessment', *Policy Options* 18: 25–28.

Stone, Richard. 1998. 'A Bold Plan to Re-create a Long-lost Siberian Ecosystem', *Science* 282, (5386): 31–34.

Stone, Richard. 1999. 'Siberian Mammoth Find Raises Hopes, Questions', *Science* 286, (5441): 876–79.

Strathern, Marilyn. 2000. *Audit Cultures: Anthropological Studies in Accountability, Ethics and the Academy.* London: Routledge.

Strong, David, and Eric Higgs. 2000. 'Borgmann's Philosophy of Technology', in *Technology and the Good Life?* ed. Eric Higgs, Andrew Light and David Strong. Chicago: University of Chicago Press, pp. 19–37.

Struzik, Ed . 1995. 'The Muskox of Banks Island', *Above & Beyond Magazine.* July 1995: 51–55.

Stubkjær, Henrik. 1998. 'Hvad laver de på Grønlands Naturinstitut?', *Sermitsiaq,* 11 September, p. 17.

Svensson, T. 1997. Th*e Sámi and Their Land. The Sámi vs. the Swedish Crown. A Study of the Legal Struggle for Improved Land Rights: the Taxed Mountains Case.* Oslo: Novus forlag.

Sveg Case lower court verdict 1996. Sveg Tingsrätt, Meddelad av Svegs Tingsrätt den 21 Februari 1996 i det så kallade Renbetesmålet (Communicated by Sveg's lower court 21 February 1996 in the so-called Reindeer Grazing Case). Dom nr. (Verdict no.) DT 12, Mål nr. (Case no.) T 88/90, T 70/91 och (and) T 85/95.

Syrovatskii, D.I. 2000. *Organizatsiia i ekonomika olenevodcheskogo proizvodstvo.* Yakutsk.

Tanner, Adrian. 1979. Bringing Home Animals: Religious Ideology and Mode of Production of the Mistassini Cree Hunters. New York: St. Martin's Press.

Tener, J.S. 1960. 'The Present Status of the Barren-ground Caribou', *Canadian Geographical Journal* 60(3): 98–105.

Tester, F.J. 1981. 'Northern Renewable Resource Managements: Socio-psychological Dimensions of Participation', *International Symposium on Renewable Resources and the Economy of the North,* ed. M.M.R. Freeman. Ottawa: Association of Canadian Universities for Northern Studies, pp. 190–97.

Tester, F.J. and P. Kulchyski. 1994. *Tammarnit (Mistakes): Inuit Relocation in the Eastern Arctic, 1939–63.* Vancouver: University of British Columbia Press.

Theberge, John B. 1981. 'Commentary: Conservation in the North – An Ecological Perspective', *Arctic.* 34(4): 281–85.

Thomas, Donald C. and James A. Schaefer. 1991. 'Wildlife Co-management Defined: The Beverly and Kaminuriak Caribou Management Board', *Rangifer* 11, special issue 7: 73–89.

Thorpe, Natasha, and S. Eyegetok. 2000a. 'The Tuktu and Nogak Project Brings Elders and Youth Together', *Native Journal* 9(7): 9.

Thorpe, Natasha, and S. Eyegetok. 2000b. 'The Tuktu and Nogak Project Elder-Youth Camp', *Ittuaqtuut* 2(2): 32–43.

Thorpe, Natasha, S. Eyegetok, N. Hakongak, and Kitikmeot Elders. 2003. *Thunder on the Tundra: Inuit Qaujimajatuqangit of the Bathurst Caribou.* Vancouver: Douglas and McIntyre.

Thorpe, Natasha L. 1998. 'The Hiukitak School of Tuktu: Collecting Inuit Ecological Knowledge of Caribou and Calving Areas Through an Elder Youth Camp', *Arctic* 51(4): 403–8.

Thorpe, Natasha L. 2000. 'Contributions of Inuit Ecological Knowledge to Understanding the Impact of Climate Change on the Bathurst Caribou Herd in the Kitikmeot Region, Nunavut'. Thesis (M.R.M.), Simon Fraser University.

Tigullaraq, J. 2000. Assistant Director, Department of Sustainable Development, Government of Nunavut: Iqaluit, NU. Personal communication to N. Thorpe.

Tobiassen, Susanne. 1998. 'Grønlandiseret forvaltning', in *Retsforhold og samfund i Grønland*, ed. Hanne Petersen and Jakob Janussen. Nuuk: Ilisimatusarfik, pp. 165–89.

Toews, Stephen. 1998. '"The place where people travel": The archaeology of Aulavik National Park, Banks Island'. Unpublished report. Western Canada Service Centre, Cultural Resource Management, Parks Canada, Winnipeg.

UNESCO. 1992. 'Editorial' *Tek talk* 1(1): 1.

Urquhart, Doug R. 1973. 'Oil Exploration and Banks Island Wildlife: A Guideline for the Preservation of Caribou, Muskox and Arctic Fox Population on Banks Island, N.W.T.' Unpublished report, N.W.T. Game Management Division. Department of Resources, Wildlife and Economic Development, Government of the Northwest Territories, Yellowknife.

Urquhart, Doug R. 1989. 'History of Research', in *People and Caribou in the Northwest Territories*, ed. E. Hall. Yellowknife: Government of the Northwest Territories, Department of Renewable Resources, pp. 95–101.

Urquhart, Doug R. 1996. 'Caribou Co-management Needs From Research: Simple Questions – Tricky Answers', *Rangifer*, special issue 9(16): 263–71.

Usher, Peter J. 1971. *The Bankslanders: Economy and Ecology of a Frontier Trapping Community.* Volume 2: Economy and Ecology. Northern Science Research Group, Department of Indian Affairs and Northern Development, Ottawa.

Usher, Peter J. 1986. *The Devolution of Wildlife Management for Wildlife Conservation in the Northwest Territories.* Ottawa: Canadian Arctic Resources Committee.

Usher, Peter J. 2000. 'Traditional Ecological Knowledge in Environmental Assessment and Management', *Arctic* 53(2): 183–93.

Ustinov, V. 1956. *Olenevodstvo na Chukotke.* Magadan: Magadanskoe knizhnoe izdatel'stvo.

Verkhovnyi Sovet SSSR. 1989. *Konstitutsiia (osnovnoi zakon) Soiuza Sovetskikh Sotsialisticheskikh Respublik.* Moscow: Iuridicheskaia Literatura.

Vestergaard, Torben. 1992. 'Industrisamfundets primitive liv', in *Livsformer og kultur. Antropologien i praktisk anvendelse*, ed. Ester Fihl and Jens Pinholt. København: Akademisk Forlag, pp. 106–14.

Vibe, Christian. 1967. 'Arctic Animals in Relation to Climatic Fluctuations', *Meddelelser om Grønland* 170(5).

Viemose, Jørgen. 1977. *Dansk kolonipolitik i Grønland.* København: Demos.

Warry, Wayne. 1998. *Unfinished Dreams. Community Healing and the Reality of Aboriginal Self-Government.* Toronto: University of Toronto Press.

Wegren, S.K. 1994. 'Yel'tsin's Decree on Land Relations: Implications for Agrarian Reform', *Post-Soviet Geography* 35(3): 166–83.

Wenzel, George. 1991. *Animal Rights, Human Rights: Ecology, Economy and Ideology in the Canadian Arctic.* London: Belhaven.

Wenzel, George. 1999. 'Traditional Ecological Knowledge and Inuit: Reflections on TEK Research and Ethics', *Arctic* 52(2): 113–24.

Wenzel, George. 2000. 'Inuit Subsistence and Hunter Support in Nunavut', in *Nunavut: Inuit Regain Control of Their Lands and Their Lives*, ed. J. Dahl, J. Hicks and P. Jull. Copenhagen: International Work Group for International Affairs. Document 102, pp. 180–93.

Wildlife Management Advisory Council. 2001. 'Species Status Reports – Muskox', www.taiga.net/wmac/researchplan/reports/muskox.html (searched 16 September).

Wilkinson, P.F. and C.C. Shank. 1975. 'Archaeological Observations in North Central Banks Island', *Arctic Anthropology* 12:104–12.

Wilkinson, P.F., C.C. Shank and D.F. Penner. 1976. 'Muskox-caribou Summer Range Relations on Banks Island, N.W.T.' *Journal of Wildlife Management* 40(1): 151–62.

Will, Richard T. 1985. 'Nineteenth Century Copper Inuit Subsistence Practices on Banks Island, N.W.T.' Unpublished PhD dissertation, Department of Anthropology, University of Alberta, Edmonton.

Willett, M. 2000. Teacher: Yellowknife. Email to N. Thorpe.

Wilson, Emma. 2002. 'Making Space for Local Voices: Local Participation in Natural Resource Management, North-eastern Sakhalin Island, the Russian Far East'. Unpublished PhD dissertation, University of Cambridge.

Wishart, Robert. 1999. 'A Preliminary Report on Caribou Hunting Practices Among Gwich'in Residing in and around Fort McPherson, NWT'. Unpublished manuscript.

Woollett, James M. 1991. 'Cultural Perceptions of Man-animal Relationships and Carcass Utilization'. Unpublished MA thesis. Department of Anthropology, University of Alberta, Edmonton.

yourYukon. 2001. 'Movements of Muskox are a Mystery'. www.taiga.net/your Yukon/col89.html (searched 17 September).

Yukon Department of Renewable Resources. 2001a. 'Yukon Wildlife at Risk'. www.renres.gov.yk.ca (searched 26 August).

Yukon Department of Renewable Resources. 2001b. 'Explore the Wild!'. Wildlife Interpretation Centres, Calendar of Special Events, September.

Zabrodin, V.A. (ed.) 1979. *Severnoe olenevodstvo*. Moscow: Kolos.

Zimov, S.A. and Chuprynin, V.I. 1991. *Ekosistemy: ustoichivost', konkurentsiia, teslenapravlennoe preobrazovanie*. Moskva: Nauka.

Index